WAKAN TANKA

On Human Origins, Spirituality and the Meaning of Life

By
JOHN BENNETT

 FriesenPress

Suite 300 - 990 Fort St
Victoria, BC, V8V 3K2
Canada

www.friesenpress.com

Copyright © 2021 by John Bennett
First Edition — 2021

All rights reserved.

No part of this publication may be reproduced in any form, or by any means, electronic or mechanical, including photocopying, recording, or any information browsing, storage, or retrieval system, without permission in writing from FriesenPress.

Image credits and licensing information can be found at the end of the book

ISBN
78-1-5255-7693-5 (Hardcover)
978-1-5255-7694-2 (Paperback)
978-1-5255-7695-9 (eBook)

Social Science, Anthropology

Distributed to the trade by The Ingram Book Company

Table of Contents

Prologue
The People on the Ledge v

PART ONE
Our Lowly Origins

Chapter One
The Age of Mammals 3

Chapter Two
The Emergence of Hominins 15

Chapter Three
The Trials of the Pleistocene 35

Chapter Four
The Sapiens 73

Chapter Five
Relying on Instinct 105

Chapter Six
The Great Rift Valley – Land of Our Birth 121

Chapter Seven
The Dawn of Human Spirituality and Culture 129

Chapter Eight
Leaving Africa 151

Chapter Nine
The Mammoth Steppe...165

Chapter Ten
The Blossoming of Human Culture 175

Chapter Eleven
Peopling of the Americas 207

PART TWO
Our Lofty Destiny

Chapter Twelve
The Remarkable Mind of Homo Sapiens Sapiens 235

Chapter Thirteen
Consciousness, Vitality, and the Soul 263

Chapter Fourteen
The Science of the Soul...281

Chapter Fifteen
The Lingering Question..291

Image Credits and Licensing..295

Bibliography
and Related Reading 301

PROLOGUE
The People on the Ledge

"It was the experience of mystery - even if mixed with fear - that engendered religion."

...Albert Einstein

IT IS MIDAFTERNOON over the savanna of East Equatorial Africa. The sun is beginning its descent towards the western horizon, and a yellow-orange hue is slowly enveloping the landscape. The remnants of a late-morning rain shower still linger, and the vibrant greens and yellows of the grasslands are glistening in the warm afternoon sun.

Snaking through the savanna is the Great Rift Valley, its massive escarpments rising sharply from the valley floor, exposing earthen layers of alternating rusty red, brown, and black. And at the base of the escarpments, the grasslands in the valley are dotted with acacia trees that support wide canopies of feathery green leaves and ancient baobabs with massive, gnarled trunks and stubby canopies. The trees cast dark shadows that are slowly creeping eastward over the landscape as the sun arcs lazily westward, across the wide, blue, African sky.

In the middle of the valley is a long narrow lake, its calm surface mirroring depth variations with a pattern of blues, greens, and browns. The lake is bordered mostly by mud banks, the vegetation having been eaten or trampled by the great herds of animals that frequently come down to the lake to drink, but here and there, stands of tall elephant grass line the water's edge, and patches of reeds and lilies extend out over the shallows towards the centre.

The valley teems with wildlife. Vast herds of ungulates—antelope, zebra, and wildebeest—graze peacefully on the grasslands. Families of elephants and giraffes feed on leaves and branches in the woodlands, while rhinoceros and warthog families graze on the tender, new grass shoots that grow near the lake.

One end of the lake is alive with a flock of pinkish-white flamingos. Hundreds wade in the shallows, filling their strangely inverted bills with silt from the lake bottom from which they filter shrimp and other crustaceans. Others stand on one leg, dozing with their bills neatly tucked beneath their wings. The air above is alive with the birds, some honking loudly as they rise from the lake with flapping wings, while others glide silently overhead in search of a landing place amongst those standing in the shallows.

At the opposite end of the lake, pelicans, egrets, herons, ducks, and small flocks of wading birds feed in the shallows, while further out in deeper water, hippopotamuses laze, completely submerged but for their nostrils and eyes, which barely break the surface.

Back from the lake, half hidden in the tall grasses and lounging in the shade of acacia and baobab trees, are the ever-present lions, leopards, cheetahs, and hyenas. Soon, they will begin their predatory forays but for now are content to lounge lazily in the afternoon heat.

Viewed from afar, the scene is calm and serene with scarcely a sound or discernible movement. There are no fences or walls marring the landscape. There are no concrete, steel, or wooden buildings, no roads or automobiles. In fact, there are no signs of human activity whatsoever, for it will be more than one hundred thousand years before humankind's great civilizations appear on the earth.

Winding in a generally east-west direction is another, smaller valley, created over the millennia by a seasonal meandering stream that emerges from the highlands to the west and snakes its way across the savanna, eventually emptying into the Great Rift Valley to the east.

Like the Great Valley, the floor of this smaller valley is mostly grassland with occasional acacia trees and bushes. At a place where the valley floor widens, a depression has formed, and during the rainy season, the stream enters the area across a broad mud flat where it slows and spreads, forming a small, marshy lake. The surface of the mud flat is a shimmering black and is heavily pocked in places with the hoof prints of the ungulates that occasionally come to drink from the lake.

It is near the end of the rainy season, however, and the stream is now reduced to a small trickle. It is only a matter of weeks before both the stream and the shallow lake are dry.

On this day, the air is still and hangs heavily in the small valley, creating a silence broken only by the occasional croak of wading birds and the muted hum of insects that flit amongst the grasses and swarm the water's edge.

Just above one end of the lake, about halfway up the escarpment, a large semicircular ledge protrudes out of the rock and earthen wall, carved out by the swirling waters of the stream many millennia ago. The floor of the ledge is smooth and flat, but at the outer edge, it drops precipitously to the valley bottom, some thirty metres below. And at one end, a beaten path leads down a winding course of switchback corners to the valley floor.

Near one end of the ledge, tucked back against the wall, stand four dome-shaped lodges, expertly built from wooden poles and wildebeest hides, sewn together with sinew and strips of animal rawhide. In front of the huts is a fire pit, consisting of a circle of stones forming a simple hearth, upon which small bits of charred meat still cling, attracting a swarm of feasting insects.

The ledge is the seasonal home to a small band of humans who, on this day, are peacefully whiling away the afternoon in light industry and relaxation. Near the centre of the ledge, five women are seated in a circle, engaged in quiet conversation as they weave baskets from reeds gathered at the lakeshore that morning. Towards the back wall of the plateau, three children are playing a game with a ball made of ungulate hide, stuffed tightly with grasses and animal hair.

Near the edge of the precipice, six men are gathered in a loose circle. One of the men is very old, and the other men are clustered around this elder, listening intently as he speaks. Two of the younger men are knapping cobbles of flint into blades while two others are lounging nearby, staring out over the valley as they listen to the elder, occasionally nodding in agreement with his words.

The band took up residence on the ledge immediately on arriving in the valley near the beginning of the rainy season and has been there for six months. They have been using the spot as a seasonal campsite for as long as any of them can remember and, by their ancestors, for many millennia before that.

The people of the band are happy and at peace with themselves and each other. Life is not easy, food is sometimes scarce, and predators are a constant worry, but this season has been a good one with plenty of food, and they are healthy and content. There are root and tuber vegetables to be foraged around the lake and berries and seeds to be harvested on the bushes and trees, and best of all, the valley has been visited by enough ungulates to keep them supplied with ample meat, hides, bone, and sinew to meet their needs.

About a month ago, however, a family of five hyenas came into the valley and took up residence in a grove of acacia trees just a short distance from the mud flat at the end of the lake. The hyenas mostly spend their days under the acacia trees, the nearly full-grown pups frolicking in the long grass and the parents sleeping in the shade. Life has been easy for them as well as they have had a good variety of small rodents and birds to prey upon around the lake, and shortly after arriving, they managed to kill a young zebra upon which they feasted for several days.

Since the hyena's arrival, however, the ungulates have mostly left the area and returned to the main herds in the Great Rift Valley to the east. The few that still occasionally graze near the lake are extremely wary, and it is virtually impossible for the hunters or the hyenas to get within killing range before being spotted. The band has been unsuccessful at killing any game since the hyenas arrived, and for almost a month, they have lived on berries, seeds, and tubers. They are hungry for meat.

As the band is passing the day lulled into a quiet languish by the peacefulness of their surroundings, one of the men suddenly lifts himself upright and points across the valley in the direction of the far side of the lake. The others follow his gaze, and soon the entire band is at full alert, all staring intently at the sight unfolding below them.

A herd of about twenty gazelles has come down the escarpment from the savanna on the far side of the valley and are cautiously moving towards the lake. The gazelles have been on the move across the savanna for several days and were attracted into the valley by the smell of green grass and fresh water. They are extremely wary creatures, however, and have stopped some distance from the lakeshore where they are standing at full alert, ears twitching and nostrils flaring. Some lift their noses skyward, searching for a scent as they sense danger but can't identify its nature or location. They are in dire need of a drink but are afraid to get any closer to the water, sensing that the danger is somehow located there. Some prance in tight circles and paw the earth with their hooves, all the time staring at the lake in obvious desperation and frustration.

The people watch intently from the ledge as this drama unfolds below them. They can see that the gazelles are wary and know that it would be almost impossible for them to get within a spear throw of the animals without frightening them off. However, from their vantage point, they can also see that the hyenas have spotted the gazelles and are stealthily working their way through the tall grass towards them.

The band grows excited, and the men huddle to quickly formulate a hunting plan. In short order, five of the men grab their spears and start down the trail that leads into the valley, staying hidden from the gazelles by crouching low behind the large rocks alongside the path. At the bottom, about fifty paces from the gazelles, they move into the tall grass and crouch low to watch and wait.

His caution overcome by thirst, a young gazelle buck suddenly breaks from the herd and begins trotting towards the mud flats and the lake. The rest of the herd stands still and watches as the buck reaches the mud flat where its pace slows as its feet sink into the soft mud. It continues to struggle forward, lifting its legs from the mud, one at a time, slowly making its way towards the water.

Suddenly, the hyenas break from the tall grass and race toward the floundering gazelle. At the last minute, the young buck catches sight of them and begins struggling desperately to gain a solid footing and escape back to the herd. But with the mud holding it back, it is too slow. The lead hyena reaches it, leaps onto its back, and knocks it onto its side into the mud. Another hyena's jaws rip at the young buck's throat, and after only a brief moment, the gazelle is dead.

The instant the hyenas make their kill, the five hunters begin to run toward them, keeping low in the grass and out of sight for as long as possible. When only a few paces away, the hunters suddenly rise up and start running directly at them, whooping and yelling and waving their spears wildly above their heads.

The startled hyenas look up at the approaching hunters and turn to face them, snarling with lowered heads and bared teeth. Undeterred, a hunter throws his spear and pierces the skin on the back of one of the hyenas. As the wounded animal screams and falls momentarily backward into the grass, the others hesitate and retreat a few paces, leaving the gazelle undefended.

While three of the hunters continue to howl and brandish their spears, the other two immediately grab the feet of the gazelle and lash them around their spears with strips of rawhide.

Hoisting the suspended gazelle onto their shoulders, the two carriers begin to move back toward the escarpment trail while the other three continue to make threatening noises and gestures toward the hyenas. As the men move, the hyenas glide alongside, looking for an opportunity to reclaim their prize. The hunters keep the hyenas at bay, however, and soon reach the base of the escarpment and the start of the trail leading up to the ledge.

Seeing where the hunters are headed, the hyenas stage a final attempt to frighten them into abandoning the gazelle. More urgently now, they circle the men, looking for an opening to recapture their prize. The hunters are resolute and respond by whooping and thrusting their spears at the hyenas, but the hyenas are undeterred and crouch and snarl in preparation for a final attack.

Suddenly, a hail of stones rains over the heads of the hunters and down upon the hyenas, striking them on their heads and backs. While the hunters were retrieving the gazelle, two of the women had worked their way down the trail to a large rock near the base. Grabbing several stones in each hand, they had climbed to the top of the rock and were waiting there with handfuls of stones and arms poised for throwing. As the hyenas and hunters reached the base of the trail, the women screamed loudly and hurled the stones at the animals.

The combination of the stones, the noise, and the brandishing spears is too much for the hyenas. Realizing that they are no match for these strange creatures, they turn as one and begin to trot back towards the lake and their grove of acacia trees. Relieved and happy, the hunters and women turn and start up the trail, carrying their prize up to the ledge where they are greeted by the rest of the band with cheers and embraces. With much pride and satisfaction, the two carriers take the pole from their shoulders and gently lay the gazelle on the ground.

The children are fascinated by the dead animal. They had been impressed at the sight of the herd and how animated and vibrant the gazelles were as they pranced nervously by the lake. Now, as they look at the dead animal lying motionless on the ground with blank, staring eyes and limp body, they are puzzled by the difference between a living and dead animal. They wonder what has happened to the animal, what has left the gazelle's body to cause such a marked transformation.

The elder then kneels down beside the animal and examines it, gently washing the mud from its fur and speaking softly, thanking it for surrendering its life to provide life-giving nourishment for his people. Then, rising and backing away from the animal, he declares it clean and calls for a celebration feast. This causes much excitement in the band as it means a night of storytelling and binging on fresh meat.

The sun is low in the sky now, and the band is anxious to get on with their feast and celebratory evening. Immediately, the adults set to work butchering the animal while the children play alongside, occasionally squeezing in beside the adults for a better look at what is happening.

Two women kneel beside the gazelle and with sharp flint blades, remove the skin from the body, carefully scraping off the meat as they pull the skin away. The hyenas hadn't caused much damage to the hide, so it will be suitable to trade for flint or tools during the next migration.

After the women have removed the skin, two men butcher the gazelle with sharp, expertly knapped, flint knives. They first cut open the belly of the animal and remove the offal, carefully placing the various organs on a grass mat that has been laid out next to the carcass. Next, they cut thick slices of meat away from bone, placing both meat and bone on the grass mat, continuing to work until the entire animal has been rendered. Useful, but inedible bones and other animal parts are piled on a second mat, and the few remaining undesired parts of the animal are thrown over the precipice to the valley floor where they will be found and carried off by nocturnal scavengers.

As the two men attend to the butchering, others prepare the cooking fire. Flint and pyrite are fetched from a shelter while wood and tinder are placed around the stone hearth of the fire pit. After just a few expert strikes of the pyrite with the flint, the tinder ignites and the fire builds gradually into a blaze. The fire will be fuelled for some time until a large bed of red-hot coals is made and the hearth rocks have heated sufficiently to cook the meat.

It is dusk by the time the butchering is complete and the hearthstones are hot enough for cooking. One by one, the band gathers around the fire, some sit on the ring of rocks, others squat or lounge on the ground. Slabs of meat, pieces of bone, and some of the entrails are placed side by side on the hearthstones until the stones are covered. The meat immediately starts to sizzle with blood and moisture seeping out of the tissue, bubbling from the searing heat.

As the meat is cooked to individual taste, the band members step up to the hearth and remove a piece of meat with a serving stick, placing it on a cooling rock for a minute or two and then picking it up in their hands to consume it. Others reach for the bones from which they will suck the delicious marrow. The feasting continues in this manner until the entire animal is consumed and the people are completely sated.

Once the meat is finished, a small amount of wood is added to the fire, just enough to keep it going and provide a source of heat and comfort as the cooler night air envelops the ledge. While they feasted, it became completely dark. Now, as the fire glows and the occasional spark rises from it, they follow each spark lazily with their eyes as it winds its way upward towards the heavens where the stars and planets shine with breathtaking number and brilliance—a sight only seen when the air has been cleansed by recent rains and there is absolute darkness. The people are content with bellies full of meat and the happiness that comes with being with friends and family around a comforting fire at the end of a perfect day.

A young man, squatting on the ground furthest away from the fire, reaches to his side and picks up a pair of carved, smooth rhythm sticks. He begins to slowly beat the sticks together, making a resonant knocking sound that catches the attention of the rest of the band. He beats the sticks in a regular rhythm for a few moments and then begins to hum in a deep baritone, "Haaaaaruuuuuuuummmmm," drawing out the sound as long as his breath holds, stopping only briefly to take another breath before resuming. The sound is primeval, haunting, yet at the same time, melodic and pleasant. Smiles appear on the faces of the people around the fire, and soon he is joined by a few of the older women who begin to hum in the same manner. The young man then begins to alter the rhythm, introducing alternating riffs and patterns with his rhythm sticks and his voice.

The simple music is captivating, enjoyed immensely by the rest of the band members. It isn't long before a man and a child stand up, hold hands, and begin to sway their bodies with the music. Soon, all the band members are doing the same with eyes closed and a look of peacefulness on their faces as they dance and sway to the rhythm of the primitive music.

The elder watches the scene with a look of peace and serenity on his face while his thoughts drift to a time long ago, and a vision of a young woman with smooth, dark skin glowing in the firelight, slowly emerges from his memory. By her side is a young boy, and as the elder savours the memory of them, his emotions overcome him, and tears stream down his face. He is suddenly filled with melancholy but also joy and thankfulness for the life he has been given and the richness of the memories he holds in his mind and heart.

After a while, the young man slows the tempo of his rhythm sticks and gradually ceases altogether while, one by one, the hummers stop and the dancers return to their seats.

Another woman, who is rather short and stout, then stands up and moves to the centre of the circle and begins to tell a story. She has a mischievous smile on her face, and the people smile with her for the stories she tells always make them laugh.

Assuming an amusing tone in her voice, she begins, "Many years ago, when I lived with a different band, there was a young woman who had not yet found a partner. But one day, this young woman noticed that she had caught the eye of a man just a few years older than herself. She could tell that this man liked her. She liked him too, but he was very shy. She tried to talk to him and play with him, but he always grew red in the face and turned away, too shy to respond to her advances. Yet when she ignored him, he would stare at her with a look of yearning and desire. This went on for two full moons."

The woman then pauses for effect, and the rest of the band leans forward in anticipation of the rest of the story. She then continues, "The young man never approached her and never came into her lodge at night. Finally, in the middle of the day, with all the band members watching, the woman walked over to the young man and, before he could turn away, reached down and grabbed him by his penis. The man was completely surprised and turned a deep shade of red. The young woman then turned and walked, penis in hand, pulling him behind her across the camp and into her lodge, with the rest of the band smirking and laughing behind them."

At this point, the woman mimics the scene, and with her hand grasping an imaginary penis behind her, she parades around the campfire before continuing, "They spent most of the afternoon together in the lodge. When they emerged, the sun was setting, and the young man was smiling broadly. He had his arm around the woman, and from that day forward, they were inseparable."

As she finishes her story, peals of laughter erupt from the people around the fire, and some are rolling on the ground, holding their sides. Knowing this woman as they do, they all know that the woman in the story is the storyteller herself, and they don't doubt its veracity. This makes it all the funnier, and it is some time before they regain their composure and the atmosphere becomes quiet once again.

Throughout the evening, several more stories are told, punctuated at times with intervals of music and dance. In the lull between the stories and music, the people are content to look at the stars and pass some time in quiet conversation or silence.

Occasionally, someone gets up and puts a piece of wood on the fire, and about midevening, the moon appears over the eastern horizon and begins its course across the night sky. The sight of the moonrise quiets the proceedings as the band members pause to watch and appreciate it.

It is one of the children that finally breaks the ensuing silence and asks, "Elder, will you tell us a story?" The others look at the old man and nod in agreement with the child. They ask him to tell them a story from long ago, before any of them were born, as such stories are the ones they love most.

And so, the elder speaks, "I'll tell you a story about something very strange that happened long ago when I was not much older than this young fellow right here," he said, patting the head of the boy sitting next to him. "It was during a migration, and our band had camped near a watering place for a few days of rest. In the afternoon of the second day, we were all sitting around, talking and sharing memories, when we noticed that something strange was happening. The sunlight was changing from bright noonday light to a dimmer, yellow-orange, and it was gradually getting darker. It was as though the sun was setting, yet it was still early in the afternoon; the sun was high in the sky, and there were no clouds covering it."

He pauses briefly to let his words settle in their minds and then continues, "We all looked up towards the sun, and it appeared that the shadow of something very large was edging slowly across its face, gradually blocking more and more of the sunlight as it moved. Eventually, it covered the entire sun, and the earth became dark, like it was late evening. A few of the night stars appeared, and the evening coolness began to settle around us. It was very frightening as no one in our band had ever seen such a thing before, and no one could explain what was happening. We children gathered close to the adults who put their arms around us as if to protect us from some terrible creature that was about to descend upon the earth. But then, after a brief period of darkness, the edge of the sun reappeared, and it began to slowly brighten again as the creature, or whatever was causing the shadow, moved on. Finally, it vanished completely, and the sunlight returned to normal as if nothing had happened."

The elder falls silent. He has never spoken of this before even though the experience has haunted him all his life. Until now, he has been reluctant to burden his people with such an unsettling story, but his time on earth is nearing an end, and he feels it important that his people should finally know about it. Perhaps knowing that it has happened in the past will make it easier for them to accept should it ever happen again.

The story was not what the members of the band had expected, and they are disturbed by it. They become agitated and much nervous chatter ensues as they try and reassure one another with possible explanations for what the elder has told them. "What could possibly be so large and powerful as to be able to block out the sun?" they ask one another. But no one can come up with a plausible explanation.

Somewhat fearfully, they huddle together, finding security in closeness to one another as they look up at the night sky. The elder's story of the darkening of the sun has deeply disturbed them, and they suddenly feel vulnerable. Even the night sky looks different—wondrous and beautiful as always, but now also frightening and foreboding.

The band falls silent as each member tries to fathom the meaning of what they have just heard, and as they contemplate the story, profound questions begin to emerge in their minds.

Gazing up at the heavens, one quietly asks, "Where do you think the earth, moon, and stars come from? Do you think there is some powerful being out there somewhere that made them?"

And another wonders, "Where do you suppose we come from? Did that powerful being make us too?"

The small band of *people* on the ledge continues to huddle together long into the night. Each ponders these questions in silence as they watch the moon and stars trace their courses across the night sky until finally, as the fire gradually grows dim, they rise, one or two at a time, and walk quietly to their lodges. They will not sleep well this night. The elder's story and the questions it has provoked have shaken them profoundly.

******

The questions pondered so long ago by that tiny band of people on the African ledge have become the enduring questions of antiquity. They are questions that have haunted humankind since the dawn of consciousness, and they continue to haunt many of us, even today.

How did the universe come to be? Where do we come from, and where do we go after death? Is there a God? What is the meaning of life?

The yearning for answers to these questions burns deep within us all and lies at the very heart of our humanity. We will never stop our seeking until we have found the answers.

PART ONE

Our Lowly Origins

CHAPTER ONE
The Age of Mammals

"I have called this principle, by which each slight variation, if useful, is preserved, by the term of Natural Selection.... And it is not the strongest of the species that survives, nor the most intelligent. It is the one that is the most adaptable to change."

...Charles Darwin

SIXTY-SIX MILLION YEARS ago, Planet Earth was a very different place than it is today. It was nearing the end of the Mesozoic geological era, and the ancient supercontinent of Pangaea had long since been torn apart by massive tectonic plate movements deep in the earth, splitting the earth's land mass into the familiar continents of today. North America and Eurasia were still joined, however, and South America and Africa had only recently separated, geologically speaking. Antarctica and Australia formed a smaller land mass in the vicinity of the South Pole, and an island next to Madagascar was being pushed northward. Some ten million years hence, the island would collide with southern Asia, forming the Indian subcontinent and the vast mountain chains of southern Asia.

The earth's climate at this time was extremely hot and humid, with annual mean global temperatures 12 to 14°C warmer than today. At such elevated temperatures across the globe, the earth was too warm for permanent snow or ice, and the water that is trapped in glaciers and the polar icepacks today, filled the seas instead. Sea levels were about one hundred metres higher than modern times, and the lower regions of North America, Eurasia, and Africa were flooded with warm shallow seas that extended well into the interior of most continental land masses. The land masses would have appeared more like clusters of islands separated by inland seas than today's continents.

With the warm seas and oceans, frequent tropical storms that carried massive amounts of evaporated water ravaged the planet, flooding the land with torrential rains. Tropical rain forests covered much of the earth's land surface, extending well into polar regions, and most of the earth's flora and fauna had evolved and adapted to living in tropical conditions.

It was the time of the dinosaurs, and the massive beasts dominated the earth's air, land, and seas. Large herds of triceratops and sauropods foraged through vast forests and woodlands of giant ferns, cycads, gingkoes, and conifers, preyed upon by fearsome meat-eating predators like the tyrannosaurus and feathered dromaeosauridae. Overhead, pterosaurs, the largest flying animals to have ever lived, soared across the landscape in constant search of prey. In the warm, shallow seas, giant, snakelike mosasaurs swam with the ancient ancestors of today's rays and sharks, and on the seabed below, sea urchins and stars were thriving. Coral reefs were beginning to form.

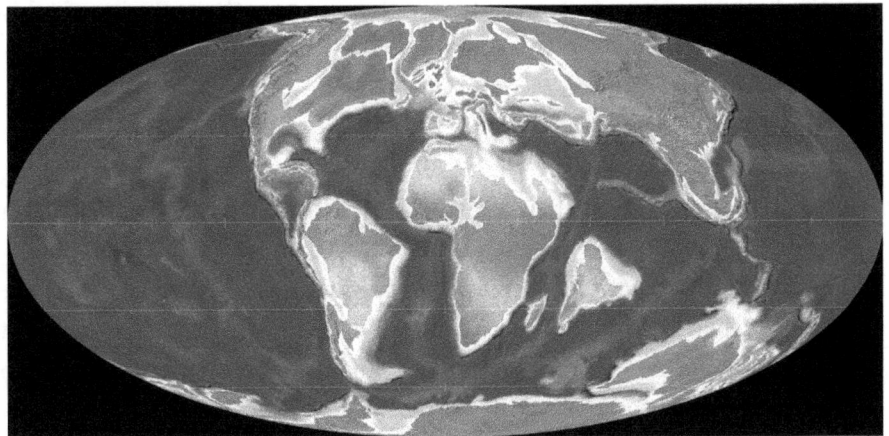

Pictorial rendition of the earth in Mollweide projection, as it might have appeared near the end of the Mesozoic era.

© *2016 Colorado Plateau Geosystems Inc.*

The earth was teeming with life at this time, and along with the dinosaurs, many smaller creatures scurried about the forests and wetlands. The ancestors of today's frogs, salamanders, turtles, crocodiles, and snakes flourished in the marshes and inland waterways. Many species of tiny shrew-like mammals foraged about the forest floor, typically living in underground burrows to escape the ever-present reptilian predators. While in the air, new species of seabirds were beginning to appear, including the ancestors of today's grebes, cormorants, pelicans, sandpipers, and loons.

It was also about this time that flowering plants appeared, and along with them, insect genera such as bees, wasps, ants, and beetles began to evolve. In the forests and woodlands, magnolia, ficus, and fragrant sassafras trees were competing with the giant ferns and conifers for sunlight.

But as the sun rose at the beginning of what would have appeared to be yet another balmy Mesozoic day, an event was unfolding in a distant region of the solar system that would mark the end of the era and forever alter the course of evolution on earth. For as the dinosaurs and other animals went about their daily routines in the swamps and coniferous forests along the east coast of present-day Mexico, an asteroid the size of a mountain was hurtling towards them at an estimated speed of over sixty thousand kilometres an hour. As they watched in incomprehension, a giant fireball streaked overhead and slammed into the earth with an unimaginably massive explosive force, estimated to have released an amount of energy equivalent to one hundred *trillion* tons of TNT.

The asteroid penetrated the earth's crust to a depth of several kilometres, gouging a crater more than 185 kilometres across and vaporizing thousands of cubic kilometres of rock. The Chicxulub Crater in the Gulf of Mexico still remains as evidence of this catastrophic event.

Within a few seconds of impact, all life within a thousand kilometre radius perished in a giant blast of thermal radiation that emanated out from the impact zone. Trees, grass, and shrubs spontaneously burst into flame, and all exposed animal life was instantly reduced to ash.

The impact generated a one-hundred-metre high tsunami that began to emanate outward across the ocean from the impact zone, wreaking havoc around coastal areas throughout the globe. At the same time, an earthquake of an intensity greater than ten on the Richter scale tore through the landscape. Enormous amounts of molten earth, rock, and debris were instantly propelled skyward, and after a few minutes, started to fall back to the earth, smothering the burning landscape with a blanket of hot grit and ash, burying the ground around the impact zone beneath metres of rubble. Droplets of airborne lava and ash exploded into the atmosphere and swirled around the globe, causing an eerie shadow to envelop the earth. An apocalyptic display of shooting stars appeared, created by the fiery debris raining back to the ground, igniting wildfires wherever it landed.

Only minutes after impact, a blast of wind would tear through the region at close to one thousand kilometres an hour, more than three times the strength of the worst hurricane ever recorded in human history. With a roar as deafening as a low-flying jet aircraft, the blast scattered debris and levelled anything that might still be standing.

The massive impact sent shock waves through the earth's crust, triggering seismic activity across the globe. Volcanoes erupted, sending more ash and debris into the atmosphere, which, in combination with the impact fallout and the smoke and ash from wildfires, enshrouded the earth in darkness. During the following weeks and months, as the dust slowly settled back to earth, the darkness turned to twilight, and the twilight to heavily clouded daylight.

For several years following the event, little sunlight reached the ground, and the earth was plunged into a prolonged period of cold, known as an impact winter. Photosynthesis was drastically reduced, flora throughout the globe withered and died, and most animal life that depended on plants for food, directly and indirectly, perished.

The event proved catastrophic for life on earth, and three-quarters of all earth's flora and fauna disappeared within a geological instant. Hardest hit were the larger terrestrial animals, and as a result, virtually the entire land-dwelling dinosaur population vanished into extinction. Following the event, only a drastically reduced number of rather small land animal species remained, and of them, most were mammals. The *Age of Reptiles* had come to an abrupt end, and the *Age of Mammals* had begun.

Most mammalian species that survived the mass extinction were small rodents scurrying about the forest floor. These first mammals lived for the most part in underground burrows, and many were nocturnal—a lifestyle that had evolved to enhance their survivability in an environment dominated by large, mostly diurnal, reptilian predators.

Gradually, however, the mammals began to adapt to the ecosystems the giant reptiles had left behind, and over the next few million years, a veritable explosion of land animal species took place. Within a fairly short evolutionary time frame, the world's land masses became densely populated with animal life of all sizes and shapes once again. But this time, mammals, as opposed to reptiles, were predominant.

For reasons not entirely understood, the high levels of carbon dioxide in the earth's atmosphere that had kept the mean global temperatures so high for millions of years began to subside about fifty million years ago. During the Eocene and Oligocene epochs, which lasted from about fifty-six to twenty-four million years ago, the earth's climate slowly, but steadily, grew cooler and drier. Sea levels began to drop, and the tropical rain forests were gradually replaced by woodlands and grasslands in some northerly areas. Global flora and fauna evolved and adapted to these changes, and many new genera of mammals appeared. The ancient ancestors of many of today's mammals, including the horse, deer, camel, elephant, rhinoceros, cat, dog, and several species of ungulates, evolved during this time.

Most of the mammals that evolved at this time were much larger than their descendants today, and some of the largest mammalian megafauna to have ever lived were roaming the earth by about thirty million years ago. As with the dinosaurs before them, the largest of these ancient megamammals were herbivores and of them, *Paraceratherium*, a genus of hornless rhinoceros, was the largest terrestrial mammal to have ever lived, weighing some fifteen to twenty tonnes and standing about five metres at the shoulder. By comparison, the largest African elephants today weigh a mere six tonnes and stand a diminutive four metres at the shoulder.

Many large carnivores appeared, the fiercest being the *Entelodonts*, an extinct family of boar-like omnivorous mammals with bulky bodies, short, slender legs, and long muzzles. Some species of the Entelodonts measured over two metres tall at the shoulder and weighed as much as four hundred kilograms, with large canine teeth and a brain about the size of an orange, a rather unfortunate combination of attributes for the other animals that were living with them at the time. The Entelodonts would have been formidable predators, and only the very largest herbivores, like the *Paraceratherium*, would have been exempt from their predation.

It was about fifty-five million years ago, near the beginning of the Eocene epoch, that the world's first prosimian primates appeared, ancient ancestors of today's lemurs, bush babies, lorises, and tarsiers. These first primates were squirrel-like in size and appearance but were beginning to show an evolutionary trend towards modern primate anatomy. They had grasping hands and feet with long, curved fingers and toes that were uniquely adapted for climbing trees and manipulating objects. Their eyes were more forward-looking, which provided them with a degree of stereoscopic vision, a capability that all primates have today.

The Messel pit, a disused quarry near the village of Messel, Germany, contains the world's most abundant fossilized remains of Eocene flora and fauna ever discovered. The site is remarkable, not only because of the large variety of fossilized plants, mammals, reptiles, fish, and insects found there, but because of the extensive preservation of structural integrity, including the fur, feathers and "skin shadows" of some species. And amongst the mammals preserved at the site are the completely intact remains of early prosimians, which have revealed many details on the anatomy, fur, and skin of the world's first primates.

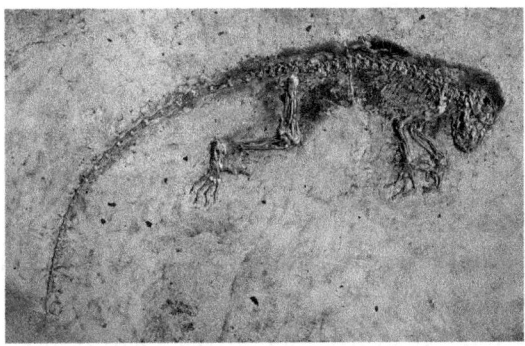

A 47 million year old prosimian fossil, *Darwinius masillae*, from Messel pit in Germany

Following their initial appearance, these early prosimians appear to have flourished, and some species began to undergo evolutionary changes that foreshadowed the simian primates or anthropoids (monkeys, apes, and humans) yet to come. Their brains and eyes began to enlarge, their snouts gradually shortened, stereoscopic vision improved, and some occasionally assumed a more upright posture.

At the base of a mammalian skull, there is a hole through which the spinal cord passes, known as the foramen magnum, which links the brain to the body's nervous system. The position of the foramen magnum is a strong indicator of the angle of the spinal column to the head and consequently can help determine whether the body is habitually held horizontally like cats and dogs or more vertically like monkeys, apes, and humans. As some of the first prosimians began to evolve towards a more anthropoid-like anatomy, the foramen magnum slowly evolved from the back of the skull towards the bottom centre, enabling them to hold their bodies erect while hopping and sitting. Ultimately, the foramen magnum would migrate completely to the bottom centre where it is found in the modern human skull.

By about forty-four million years ago, the first ancestors of apes, monkeys, and humans were evolving from the prosimians, and many primitive species began to appear in the African rain forest. These first simians resembled today's monkeys and were of various sizes, ranging from that of a fat squirrel to about the size of a small dog. Compared to their prosimian cousins, they had fewer teeth, less prognathic snouts, larger brains, and more forward-looking eyes with good stereoscopic vision. These and other anatomical features suggest that these early simians were becoming mostly diurnal tree-dwellers and practiced a lifestyle much like that of today's monkeys, spending their days travelling through the rain forest canopy in search of fruit, seeds, and insects and their nights safely nestled on a favourite tree branch.

During the Oligocene and Miocene epochs, geological time periods that lasted from approximately thirty-four to twenty-three and twenty-three to a little more than five million years ago respectively, the earth's climate continued its cooling and drying trend. The resulting regional climate shifts affected the evolutionary direction of most of the earth's plant and animal species.

The Oligocene was also an epoch of major geological change. By the beginning of the epoch, tectonic plate movements had completely separated North America and Europe, forming the distinct continents they are today. In east Africa, a two-thousand kilometre volcanically active fault zone had developed as tectonic plates deep underground were diverging, forming the Great Rift Valley. These plate movements also forced the lands to the west of the rift to lift, giving rise to the east African highlands and mountains. Today, these mountain ranges are the location of some of Africa's highest mountains, including Mount Kilimanjaro.

These massive geological changes, in combination with the moderating of the earth's global climate, trended local weather patterns even further toward cooler, drier conditions. In Africa, the rain forests retreated from the Sahara and east Africa regions, giving way to new ecosystems, characterized by sparse woodlands, open grasslands, and desert. It was at this time that the vast savanna region of east equatorial Africa began to appear.

Around the beginning of the Miocene epoch, some twenty million years ago, the first ungulates or hoofed mammals were appearing on the African savanna. Until this time, ungulates were generally smaller than today and lived in the rain forest on a diet of fruit and leaves. With the onset of environmental change, larger ungulates that could thrive on a diet of grasses alone and with the ability to travel long distances in search of food and water evolved. Descendants of these ancient grazing animals form the vast herds of antelope, zebra, and wildebeest living on the African savanna today.

As the rain forest retreated, most species of primates retreated with it and evolved into the monkeys found throughout the tropical rain forests today. But some primate species opted to leave the safety of the rain forest canopy and adopt a lifestyle that exploited the new food and other resources the new open ecosystems afforded.

Over time, while still retaining their ability to climb trees and forage in the canopy, new ape-like primates began to evolve with hybrid anatomies that allowed them to flourish in a mixed habitat of rain forest, sparse woodlands, and grasslands. Gradually, the first hominids, common ancestors of today's great apes—gorillas, orangutans, chimpanzees, bonobos, and humans—began to evolve.

The evolutionary transition from monkeys to apes took place slowly and in concert with the pace of environmental transformation that was occurring as a result of the gradually cooling climate. These early monkey/ape transitional species appear to have been remarkably successful at adapting to the new mixed habitat environment, however, as their remains are present in large numbers in the fossil record toward the end of the Oligocene epoch.

The early monkey/ape transitional anthropoids had larger, stronger bodies than their monkey ancestors, enabling them to better defend themselves against ground-dwelling predators, and they had larger brains, making them more resourceful at finding food. While all monkeys have a prehensile tail, which is of considerable advantage for creatures living primarily in an arboreal environment, the new ape-like species gradually became tailless as they spent more time foraging on the forest floor, and the survival advantage of a tail diminished. As time went on, monkeys remained physically adapted for a life in the forest canopy, and the monkey-ape transitional species evolved characteristics more suited to a combined arboreal and terrestrial lifestyle.

One of the first of the monkey-ape transitional primates was *Proconsul africanus*, the remains of which were first discovered in western Kenya by the British paleoanthropologist Dr. Mary Leakey in 1948. Since Dr. Leakey's initial discovery, the remains of at least four species of the genus Proconsul have been unearthed, all in east Africa and ranging in size between ten and thirty-eight kilograms.

Proconsul's head was distinctly apelike, with a less prognathic face, a smaller snout, and a somewhat larger braincase than a monkey. Their hands and feet had apelike flexibility, but other aspects of its anatomy suggest it habitually maintained a horizontal posture and walked through the tree canopy along the top of branches like its monkey contemporaries. By contrast, all the great apes today move through trees while suspending themselves below the branches.

A representation of *Proconsul africanus* based on a partially reconstructed skeleton on display in the Natural History Museum, Paris, France.

Image © Nobu Tamura

The Proconsul genus was enormously successful, living throughout east Africa from twenty-three to fourteen million years ago.

Following Proconsul, the trend towards more apelike anatomy continued with the appearance of several monkey-ape transitional genera throughout southern Eurasia and Africa, perhaps the most notable of which were members of the genus *Dryopithecus*.

The Dryopithecines were relatively small primates that lived primarily throughout southern France and Spain between 12.5 and 11 million years ago. Their faces were similar to that of a modern gorilla, and males appear to have been much larger, with longer canines, than the females, much like today's gorillas. A typical male Dryopithecine, however, is estimated to have weighed only about forty-four kilograms, which is much smaller than a modern male mountain gorilla, which can weigh as much as two hundred kilograms.

An artist's rendering of *Dryopithecus*.

At the time of Dryopithecus' earthly tenure, the climate of Europe was temperate with significant seasonal fluctuations in temperature, creating colder winters and frequent harsh living conditions. Dryopithecus would have had to adapt to such conditions with a heavier fur coat and an expanded diet that changed with the seasons, unlike most other primates before or since their time. The continued cooling of the European climate is also suspected to have been a factor in the disappearance of Dryopithecus from the fossil record by about eleven million years ago.

Of all the monkey-ape transitional genera that have been discovered to date, Dryopithecus is considered by many to have been the first true hominid, or *Great Ape*, and the most likely common ancestor of modern apes and humans.

The first Dryopithecus fossils were discovered in the French Pyrenees by the anthropologist Édouard Lartet in 1856, about three years prior to Charles Darwin's publication of his *Origin of Species*, and the discovery was an influential factor in Darwin's development of his theory of evolution. Darwin referenced Dryopithecus as a potential ancestor of modern humans in his *Descent of Man*, which was published in 1871.

By comparing the genomes of extant great apes, geneticists have determined that the last common great ape ancestor lived ten to twelve million years ago, after which time further speciation began to occur, ultimately leading to the appearance of gorillas, chimpanzees, and hominins (humans) as separate species. The gorilla genus first appeared approximately ten million years ago, followed by the chimpanzee-human split, somewhere between five and seven million years ago. Chimpanzees are, therefore, considered to be our closest living relative.

CHAPTER TWO

The Emergence of Hominins

"Scientists can't seem to agree on just how long we humans have been on this earth but I think we can all agree that it has been long enough for us to know better."

...Anonymous

BETWEEN TEN AND five million years ago, the earth's annual average temperature continued to fall, and the east African climate steadily trended towards cooler, drier conditions. Forests continued to shrink while open grasslands and savanna expanded. Most primate species retreated with the rain forests, but for the few that could adapt to the changing conditions, the grasslands and savanna afforded new lifestyle opportunities that included an abundant supply of meat.

On open grasslands, food is much more widely dispersed, and successful hunting and foraging requires much greater mobility. As early savanna dwellers developed a taste for meat, they required a faster method of locomotion to catch ungulates that had evolved into swift runners themselves. Seasŏnal climate variations also demanded that all animals living on the savanna migrate over long distances at various times of the year in search of more hospitable environments.

It was these food opportunities and physical demands of life on the savanna that drove evolutionary changes in hominid anatomy. While the gorilla and chimpanzee remained rain forest and woodland dwellers, and their descendants remain so today, our human ancestors or hominins, began to spend more time in the grasslands and on the open savanna, and their anatomy began to evolve accordingly. This new evolutionary path resulted in major physical changes and adaptations that would ultimately lead to the anatomy and physiology of we modern humans.

A crucial step along this new evolutionary path toward human anatomy was the development of a fully erect posture and bipedalism. Walking upright was a huge survival advantage on the savanna as it allowed for continual surveillance over further distances, more efficient walking and running, and the endurance to migrate to new territories as seasonal climate changes dictated. Studies have shown that human bipedal walking consumes only 25 percent of the energy consumed by other primate quadrupeds, and it is likely that this more energy efficient method of locomotion was a major factor driving the evolutionary transition to bipedalism.

While it may seem an easy exercise to list those traits that define human anatomy as distinct from other great apes, scientists do not all agree on the definition, especially when it comes to deciding which archaic primates should be considered part of the human family tree and which should not. This difficulty arises because the transition from our last common ape-human ancestor to anatomically modern humans took place in rather small steps. Such familiar traits as our large brains, shorter forelimbs, flexible hands, erect posture, and bipedalism developed gradually, and many *transitional* species that possessed some of these traits to a greater or lesser degree came and went. Being anatomically human, it seems, is a subjective determination as it is a matter of degree to which certain anatomical and morphological traits are present.

We humans possess by far the largest brain of all extant primates, about three times larger than chimpanzees, and much larger than would be expected of a mammal with our body size. We are also the only extant primate that is fully bipedal. And so, amongst all the differences between human anatomy and that of other primates, scientists are in general agreement that the two key defining characteristics are a large brain and full bipedalism.

Based on the paleontological evidence, bipedalism was the first of these anatomical features to evolve, having appeared some four to five million years prior to the appearance of brains as large as that of the average modern human. Hence, evidence of bipedal locomotion is an important anatomical trait that paleontologists look for when fossilized remains are analyzed in the context of the human family.

The last common chimpanzee-human ancestors, estimated to have lived about seven million years ago, are believed to have been anatomically similar to today's chimpanzees. They were undoubtedly quadrupeds with a physical anatomy that was well adapted for both tree climbing and foraging over short distances on the forest floor. In a forest, full bipedalism offered no particular survival advantage, and when these ancient primates occasionally walked on their hind legs, their gait would have been awkward, much like that of today's great apes.

As more woodlands and grasslands replaced the rain forest, food resources became more scattered, and the early primates that chose to live in these new ecosystems, for whatever reason, found themselves travelling greater distances as they hunted and foraged for food. Under such conditions, selective survival stresses favoured bipedalism, and primate anatomy began to evolve accordingly.

A comparison of human and chimpanzee anatomy provides much insight into the rather large number of evolutionary changes—some substantive, some extremely subtle—that were required to transform from a quadrupedal chimpanzee-like species to a fully biped hominin. The transition was an intricate process, requiring many precisely sequenced or simultaneous alterations to the arrangement, shape, and size of almost every part of primate anatomy, from head to toe, as it were. The fact that it happened so completely and over such a relatively short period of time, evolutionarily speaking, is testimony to the astonishing power and responsiveness of the evolutionary process.

Starting with the feet, larger heels were required to bear the increased amount of weight that had to be supported by two feet instead of four. No longer required for grasping, toe size shrank and the opposable big toe that enabled quadrupeds to swing with ease through the rain forest canopy reduced in size and relocated to align with the other four toes of the foot, making bipedal walking more comfortable and energy efficient. The foot itself arched to facilitate weight transference while walking, improving energy efficiency even more.

An increase in leg length and a thickening of the femur occurred as bipedalism demanded a change in the way leg muscles functioned during an upright gait. In bipedal walking, the *push* for moving forward comes from the leg muscles acting at the ankle. A longer leg allows the use of the natural swing of the limb when walking, so humans do not need to use muscles to swing the other leg forward for the next step the way all great apes do. We humans walk with our upper body more or less stationary, while chimpanzees and gorillas must lift and rotate their entire upper body with each step.

Knee joints enlarged to better support an increased amount of body weight, and a change in the pattern of the knee joint enabled *double knee action*, decreasing energy loss by minimizing vertical movement of the body's centre of gravity when walking. Humans walk with their knees kept straight and their thighs bent inward, so the knees are almost directly under the body, rather than out to the side, as is the case with other apes. This type of gait also aids balance.

Chimpanzees can stand on their hind limbs, but their femurs are not adapted for a prolonged upright posture or bipedal walking, and they cannot stand for long periods of time without getting tired. Quadrupeds have vertical femurs, while bipeds have femurs that are slightly angled medially from the hip to the knee. This adaptation allows our knees to be closer together and under the body's centre of gravity, permitting the locking of knees and standing up straight for long periods of time without much effort from the muscles. The gluteus maximus, one of the largest muscles in humans and much larger than in all other apes, assumes a major role in walking and running. When bipedal primates run, their upright posture tends to flex forward as each foot strikes the ground, creating a forward momentum. The gluteus muscle helps to prevent the upper trunk of the body from pitching too far forward and falling over.

Compared with chimpanzees, humans have larger, broader, and shallower hip joints that are better able to support the greater amount of body weight passing through them. This and other associated anatomical changes to the pelvis and spinal column provided a stable base for supporting and balancing the torso while walking upright.

As a consequence of bipedalism, forelimbs and hands were no longer needed for locomotion, and they evolved in a manner that is optimized for carrying, holding, and manipulating objects with great precision. This results in decreased strength in the forelimbs relative to body size, but having long hind limbs and short forelimbs facilitated an upright posture, while chimpanzees have maintained the longer arms of their ancient ancestors to facilitate quadrupedal walking and swinging on tree branches.

The human skull is balanced on the vertebral column, and the foramen magnum is located under the skull, which puts much of the weight of the head behind the spine. Furthermore, the flat human face helps to maintain balance on the occipital condyles, the prominent bones around the foramen magnum. Because of this, the erect position of the head is possible without the prominent bone ridges over the eyes that provide the strong muscular attachments found in chimpanzees, and as a result, the muscles of the forehead in humans are used only for facial expressions.

Because of the large number of anatomical alterations required to transition from quadrupedal to bipedal locomotion, scientists are able to look for clues to the transition in almost every part of fossilized remains, be it a leg bone, partial skull, or even a jaw bone. Complete fossilized skeletons are extremely rare and, more typically, only a small number of bone fragments are found. Yet it is often still possible to assess the degree to which a particular specimen was bipedal because of the large number of anatomical differences, in almost all parts of anatomy, between quadrupedal and bipedal species.

The transition to bipedalism amongst the early hominids appears to have begun about seven million years ago with the appearance of *Sahelanthropus tchadensis* that lived sometime between seven and six million years ago in the Sahel region of present-day Chad. Existing fossils include only a relatively small cranium, five pieces of jaw, and the braincase, which has a volume of approximately 350 cubic centimetres. The braincase has been named *Toumaï*, which means *hope of life* in the local Daza language of Chad.

Tchadensis had features that resemble a blend of chimpanzee and human anatomical characteristics, but the extent to which the species was bipedal cannot be determined with certainty. The foramen magnum's location near the bottom centre of the skull is, however, indicative of an upright posture and likely at least part-time bipedalism.

The remains of another early, potentially bipedal hominid, *Orrorin tugenensis*, were unearthed in the Tugen Hills of central Kenya, east Africa, in 2001. More than a dozen fossil fragments, which exhibit a novel combination of ape and human traits, dating between about 6.2 and 6.0 million years old have been found at the site. Tugenensis was approximately the size of a chimpanzee but had smaller teeth with thick enamel, similar to a human's. The size and shape of the upper femur indicates that *Orrorin tugenensis* probably walked upright on two legs for at least part of the time, but how routinely it did so can only be speculated.

The Ardipithecines

The oldest fossilized primate remains yet discovered that show undisputed signs of bipedalism are members of the genus *Ardipithecus*, which some paleoanthropologists consider to be the first true hominin and the earliest members of the human family tree.

The Ardipithecines inhabited a mixed habitat of marshes, woodlands, and grasslands, and their anatomical features indicate that they would have been equally adept at climbing trees to access fruit and seeds and walking upright while foraging around the forest floor. While they likely walked upright on two legs for greater distances than their ape relatives, they still possessed opposable toes and could not have walked comfortably for long distances. They would also have been pretty much incapable of running, so their bipedalism would have been of little value when it came to escaping predators.

Perhaps the most notable of the Ardipthecines was *Ardipithecus ramidus* that lived in the Awash River valley of Ethiopia between 4.5 and 3.0 million years ago. Many fossilized remains of the species have been uncovered, including one of the most complete early hominin specimens ever found: a female with over one hundred different bone fragments, including a nearly complete skull, teeth, pelvis, hands, and feet. When alive, the individual would have weighed about fifty kilograms and stood about 110 centimetres tall, and although she was clearly at least a part-time biped, she still had opposable big toes and thumbs. Her remains have been nicknamed "Ardi" by the paleontologists who found her, and she lived about 4.4 million years ago.

Ardi was a hybrid hominin in that the size and shape of her legs, feet, and pelvis suggest that she walked upright when on the ground like a hominin, but her four grasping hands and feet would have enabled her to move through the tree canopy with distinctly apelike prowess. She had opposable big toes so she would not have been capable of walking upright for long periods of time, but her wrist bones were more flexible and her palm bones were shorter than an ape, indicating that Ardi most likely did not walk on her knuckles like modern apes except perhaps when moving along tree branches. In short, Ardi would have appeared apelike when foraging in the tree canopies but somewhat humanlike when on the ground.

An artist's rendering of *Ardipithecus ramidus*.

© Jay H. Matternes

Ardi had a small braincase, measuring between 300 and 350 cubic centimetres, slightly smaller than a modern chimpanzee and roughly a quarter the size of today's average human.

The teeth of *Ardipithecus ramidus* lacked the specialization of other apes, indicating that they were a generalized omnivore with a diet that included soft plants, fruit, and meat as opposed to fibrous plant material (e.g., roots, tubers) or hard, abrasive food. Given their relatively small size and lack of weapons of any kind, their meat consumption would have been limited to small rodents or carcass remains left behind by more capable predators.

Some paleoanthropologists speculate that sexual dimorphism was much reduced with the Ardipithecines, based on various anatomical comparisons. The upper canine teeth, in particular, were smaller than a chimpanzee's, and those of the males were not distinctly different from that of females, which has prompted some researchers to speculate that male-to-male conflict would have been less prevalent than that observed in modern chimpanzees. This reduced conflict may have led to increased pair-bonding and parental care for offspring. In essence, the nuclear family unit may have begun to appear as early as four million years ago amongst the Ardipithecines.

The Australopithecines

About 3.7 million years ago, in the Laetoli region of northern Tanzania, three hominins walked along a woodland path covered with a thick layer of powdery volcanic ash, the fallout from the recent eruption of a nearby volcano. It was the rainy season, and a gentle rain shower had moistened the ash, so as they walked, they left well-defined footprints in the soft volcanic mud.

A short time after their passage, further volcanic eruptions blanketed the region with additional ash deposits, and over time, the ash hardened into tuff, and the tuff was buried as vegetation regrew and soil accumulated. There the footprints remained, exquisitely preserved for millions of years until erosion finally brought them back to the surface and they were discovered by Dr. Mary Leakey in 1976.

Dr. Leakey and her team eventually excavated a twenty-seven-metre trail of prints that contained a total of some seventy footprint impressions. It was determined that the footprints were produced by three individuals walking in two tracks, with one track of small prints beside a second track of larger ones. A third set of footprints of medium length are superimposed on top of the larger prints suggesting that a third individual was walking in the footsteps of the larger one.

The footprints are remarkably well preserved and the result of an extremely rare, almost ideal combination of geophysical events. The mud of volcanic ash made near perfect casts of each print and having been hidden underground in a geologically stable stretch of tuff rock for millions of years, they survived to the present day virtually intact.

Well-preserved footprints reveal a great deal about the creature that made them. The overall size and shape of the foot and toes, the extent of the arch, the depth of penetration of different parts of the foot, and the stride length all serve to reveal anatomical details about the individual that left them.

The most striking aspect of the Laetoli footprints is that they are virtually indistinguishable from those of a habitually barefoot, modern human. The big toe, in particular, is non-opposable and in line with the other toes, indicative of full-time bipedal locomotion. The foot was arched, and the footprint impression indicates a leisurely stride, with the heel striking first, followed by a weight transfer to the ball of the foot before pushing off the toes, all similar to the modern human walking stride. While taller than the Ardipithecines, the stride length indicates that the hominins that left them were still quite short, between 120 and 140 centimetres tall. The average height of humans today is approximately 160 centimetres for females and 170 centimetres for males.

Paleoanthropologists remain passionately divided over the likely relationship of the three individuals that made the footprints. Some argue that there is no basis to conclude the individuals were related or even that they walked the path at the same time. Others point out that the smaller footprint trail bears penetration variations suggesting the smaller individual was burdened on one side. From this, they speculate that the smaller tracks may be that of a female carrying an infant on her hip, and that the group may have been a nuclear family. The truth can never be known, but the possibilities are indeed intriguing.

Paleoanthropologists do agree, however, that the Laetoli footprints were left by an Australopithecine.

The Australopithecines lived in eastern, central, and southern Africa from approximately 4.2 to 2.2 million years ago, and the genus is considered by most paleoanthropologists to be the immediate ancestor of the genus *Homo*. They were bipeds with a general anatomy that reflected a combination of both ape and human-like features. They were still diminutive, standing 120 to 140 centimetres tall with a larger brain volume than the Ardipithecines but still only about 35 percent of that of the average modern human. Given their small stature and lack of hunting or butchering tools, the Australopithecine diet was mostly vegetarian with only the occasional meal of meat from smaller rodents and prosimians or of the remains of larger kills left behind by other predators. It is also likely that the Australopithecines were preyed upon by the large cats and hyenas that lived in the area.

All Australopithecines were bipedal with non-opposable, in-line toes and arched feet, with leg and torso anatomy that had evolved to support an erect posture. Hence they would have been capable of walking on more open terrain for fairly long periods of time. But they also retained longer forelimbs with long, curved fingers and thumbs, which enabled them to climb trees and move through the forest canopy more easily than we modern humans. Compared with their immediate ancestors, the Ardipithecines, who had opposable toes and spent most of their time in trees, the Australopithecines showed a marked evolvement towards full-time bipedalism, indicating that more of their time was spent wandering over greater distances as they foraged throughout their mixed habitat of woodlands, wetlands, grasslands, and occasionally, the open savanna.

The best known of the Australopithecines is *Australopithecus afarensis,* which lived throughout the Great Rift Valley in east Africa between 3.8 and 2.9 million years ago. The species was evolutionarily very successful, having survived for over nine hundred thousand years, more than four times longer than we humans have been on the earth. To date, the remains of over three hundred individuals have been found at various sites along the Great Rift Valley of Ethiopia, Kenya, and Tanzania, and it was *A. afarensis* individuals that left the footprints at Laetoli.

Australopithecus afarensis was slenderly built, with males averaging about 130 centimetres in height and females somewhat shorter. They had apelike face proportions with a flat nose and strongly projecting lower jaw; a small brain, about one-third the size of the average human today; and long, strong arms with curved fingers adapted for climbing trees. They also had small canine teeth like all other early humans and were habitually bipedal. These evolutionary adaptations for living both in the trees and on the ground were no doubt a major contributing factor to their survival for almost a million years, during which their environment cooled and dried, and forests and woodlands were replaced with grasslands and savanna.

An artist's rendering of *Australopithecus afarensis*.

© *2005 by Encyclopædia Britannica, Inc.*

The Australopithecines coexisted with many types of predators, and predation would have been the most common cause of death. The social organization of groups would, therefore, have been crucial in providing some degree of defence against predators. In this regard, a remarkable fossil find in the Lower Awash Valley in Ethiopia may offer some clues as to the social organization of bands of Australopithecines.

The find, sometimes referred to as the "First Family," is a collection of 3.2-million-year-old remains of at least seventeen *A. afarensis* individuals, consisting of both male and female adults, three adolescents, and five young children, constituting what appears to have been an extended family. The evidence indicates that the entire group died at the same time, but the cause of their sudden demise is the subject of much speculation, with theories including drowning in a flash flood, a multi-carnivore predatory slaughter or food poisoning.

The most famous *A. afarensis* fossil found to date is the partial skeleton of a female discovered in the Lower Awash Valley in 1974 by paleoanthropologist Donald Johanson. The Beatles tune "Lucy in the Sky with Diamonds" was popular at the time of the discovery, and Johanson chose the name Lucy for his find. Lucy is also known as *Dinkinesh*, which in the local Amharic language means, "You are marvellous."

Several hundred of Lucy's fossilized bone fragments were found, dated to approximately 3.2 million years ago, representing 40 percent of her entire skeleton and enough to assemble an accurate reconstruction of her complete anatomy. As a result, much of what is known of *Australopithecus afarensis* has been determined from Lucy's fossilized remains.

The youngest Australopithecines, *Australopithecus africanus* and *Australopithecus garhi*, are believed to be the immediate lineal ancestors of the genus *Homo*. Africanus was first discovered in 1924 in a limestone cave system near the town of Tuang, South Africa, and the remains are sometimes referred to as the Tuang child. It was the first of the Australopithecines to be found, and its discovery fiercely divided the scientific community as, prior to that time, paleoanthropologists generally believed that humans had their origins in Eurasia. After much turmoil and heated debate that spanned almost thirty-five years, the significance of the find was finally acknowledged, and Africa was accepted as the humanity's birthplace. The complex of limestone caves in South Africa where the remains were found is today known as the Cradle of Humankind.

Australopithecus africanus lived between 3.0 and 2.2 million years ago and was anatomically very close to afarensis, with a similar integration of human and apelike features. Its cranium was somewhat rounder than a typical ape's, with a somewhat larger brain and smaller teeth. It also retained some apelike features, including arms that were slightly longer than the legs and a strongly sloping face with a pronounced jaw. Like all Australopithecines, the pelvis, femur, and foot bones of africanus indicate that it was a biped, but its shoulder and hand bones were still adapted for climbing.

A cranium, a few skull fragments, and a partial skeleton of what appears to be the youngest lineal ancestor of the genus *Homo*, *Australopithecus garhi*, were found in the Middle Awash Valley region of Ethiopia, dated to about 2.5 million years ago. The fragmentary skeleton indicates a longer femur, suggesting that garhi may have been taller than the average Australopithecine with a longer stride, although the long, powerful arms and grasping fingers characteristic of an Australopithecine are still present.

The fossilized remains of *Australopithecus garhi* were also found with antelope bones that bear cut marks, apparently from primitive, shaped, stone tools that were also found with the fossilized skeletal remains. It appears that *Australopithecus garhi* may have been the first hominin to use stone tools and enjoy more meat and bone marrow as part of their regular diet.

South Africa's candidate for the position of the youngest lineal ancestor of the genus *Homo* is *Australopithecus sediba*, whose two-million-year-old remains were first discovered in dolomitic hills north of Johannesburg, South Africa. The remains were those of a juvenile male and an adult female who appear to have fallen to their death down a deep, vertical shaft in the dolomite rock formation. The remains show many similarities with *A. africanus* but also a few key differences that trend toward more modern human like features. *A. sediba* had smaller molars and less pronounced cheekbones than their Australopithecine relatives with longer legs, and a more humanlike pelvis, hip, knees, and ankles, all indications that *A. sediba* was a habitual biped. *A. sediba*'s braincase size was roughly 420 to 450 cubic centimetres, roughly in line with other Australopithecines, but it had a surprisingly modern hand, with a long thumb and short fingers, suggesting a precision grip suitable for toolmaking, although to date, no evidence of toolmaking has been found associated with sediba's remains.

The Australopithecines were the first truly bipedal hominins, and they appear to have used their increased mobility to advantage as their remains are widely distributed along the Great Rift Valley from the Afar Triangle of northern Ethiopia to the limestone caves of South Africa.

The Beginning of the Stone Age

"The reasonable man adapts himself to the world; the unreasonable one persists in trying to adapt the world to himself. Therefore all progress depends on the unreasonable man."

...George Bernard Shaw

The development of stone tool technology is regarded by many as a hallmark of the genus *Homo*, and an important distinction in this regard is *opportunistic* tool use versus *premeditated* tool fabrication. While other animal species use rudimentary tools to assist them in obtaining food, they typically do so opportunistically; that is, their tools are naturally formed objects, searched out as required and discarded after use. Only humans, it seems, practice premeditated tool construction, modifying natural objects to make improved tools and keeping them handy for future use. The point in evolution at which early hominins transitioned from opportunistic to premeditated tool use is an important sign of increased intelligence and one of the first indications of human cultural development.

To understand how tool use likely developed in early hominins, scientists have studied tool use by our closest living relative, the chimpanzee. It has been observed that chimpanzee communities often use stones as hammers to crack nuts, and wood or stone as weapons in hunting small animals. The most sophisticated chimpanzee tools are small, slender tree branches from which they strip off the leaves and use as probes for extracting termites and ants from inside their nests. Chimpanzees have also been observed using sticks as short thrusting spears to kill galagos, a small nocturnal prosimian, which they find sleeping in holes and crevices of trees.

But while it can be argued that fashioning a stick by removing branches and leaves is a form of tool fabrication, such tools are typically discarded after use, and no chimpanzee has been observed fashioning such tools before they were required or keeping them for very long after they were no longer needed. More importantly, no other animal, including chimpanzees, has been observed knapping stone, a much more difficult and advanced form of tool fabrication but which results in more effective, longer-lasting tools.

It is likely that early Australopithecines initially acquired tools opportunistically in much the same way as modern chimpanzees, although there is some evidence that *A. garhi* may have begun the transition to primitive stone tool fabrication.

Premeditated tool fabrication by ancient hominins likely came about in concert with the increase of meat in their diet. Early hominins ate their meat raw, and they lacked the large canine teeth and sharp claws necessary to tear meat off the carcass of a large animal. Stone tools were necessary to scrape hides away from meat, to separate meat from bone, and to crush bone for the extraction of marrow.

Currently, the oldest archaeological evidence of tool use by ancient human ancestors are the 3.4-million-year-old fossilized animal bones bearing scraping and percussion marks found in the Lower Awash Valley in Ethiopia. These first stone butchering tools would have been naturally formed objects such as split or broken rocks; nevertheless, it indicates that stones were used as tools very early in hominin evolution, and scientists consider this discovery as the beginning of the Palaeolithic or Stone Age. The use of stone as the primary material for toolmaking lasted more than three million years and is generally considered to have ended when humans discovered metalworking at the beginning of the Bronze age, some five to seven thousand years ago. Nevertheless, there are indigenous people living in remote regions of the earth today who fashion and use stone as their primary material for fashioning hunting weapons and butchering tools.

After a time, early hominins noticed that tools worked more efficiently if the cutting edge was sharp, and eventually the technique of *knapping* was discovered, the process whereby one rock is struck with a second one to chip away flakes and form a sharper edge. The technique appears to have been adopted about 2.5 million years ago, based on the discovery of the remains of the world's first confirmed stone knapper by paleoanthropologists Mary and Louis Leakey at Olduvai Gorge, Tanzania, in 1960. The Leakeys named the discovery *Homo habilis*, which literally means *handyman*, a name chosen because of the large quantity of hand-crafted stone tools found in close proximity to the fossilized remains.

An artist's rendering of *Homo habilis*.

© 2005 by Encyclopædia Britannica, Inc.

Homo habilis roamed the Rift Valley of East Equatorial Africa from approximately 2.4 to 1.4 million years ago and had anatomical characteristics closely resembling the Australopithecines but trending further in the direction of modern humans. Like the Australopithecines, habilis was diminutive, standing roughly 130 centimetres in height and weighing about thirty-two kilograms. Their forearms were longer than their legs, and their hands retained the long, curved fingers suited to climbing trees and foraging in the canopy. Their braincase, however, with a volume of about 640 cubic centimetres, was 50 percent larger than that of the average Australopithecine, but it was still slightly less than one-half the size of the average human today. Their face and teeth were more humanlike, smaller and less prominent than earlier Australopithecines.

The tools associated with the remains of *Homo habilis* represent the earliest period of stone-knapping technology, known as the Oldowan tradition, a name derived from their initial place of discovery, Olduvai Gorge. It appears habilis carefully selected hard, water-worn flint cobblestones, likely from nearby streams, and fashioned tools from them by using a second hammer rock to knock flakes off one end, thereby producing a sharp edge. The result, known as a core tool, was a cleaver-like implement with a sharp, jagged edge at one end and a smooth, rounded handle at the other, which fit comfortably in the hand. These core tools are the Palaeolithic equivalent of the ulu made in modern times by the northern Inuit people and were likely used for the same task, skinning and butchering animals. The pieces of stone removed during the knapping process, known as *flake* tools, were used as knives for finer tasks like slicing meat away from bone and possibly fashioning implements out of wood and other perishable material, the remains of which, unfortunately, don't survive in the fossil record. Evidence of flake tool use to slice meat away from bone can be seen in cut marks that are still visible on fossilized bones of animals harvested for food.

A stone chopper fashioned in the Oldowan tradition.

There is no evidence that *Homo habilis* made weapons for killing large animals, and they most likely scavenged meat from the kills abandoned by other large predators. At times, they would have hunted monkeys and other small game, but fruits, nuts, and vegetables were the mainstay of their diet.

With a small stature and still somewhat limited ambulatory speed and agility, *Homo habilis* was an easy prey for the large predators with whom they coexisted. There is ample fossil evidence to indicate that, at that time, hominins were something of a staple in the diet of large predatory animals such as the leopard and *Dinofelis*, a now extinct scimitar-toothed cat the size of a jaguar.

Because of their larger brain and toolmaking capabilities, habilis is classified by many paleoanthropologists as the first species of the genus *Homo*, but others point to their small stature and hybrid anatomical features and label them as Australopithecine. Regardless of taxonomical classification, *Homo habilis* or *Australopithecus habilis* was, at the very least, a transitional species between the Australopithecines and members of the genus *Homo* and possibly a direct, lineal ancestor of we modern humans.

Another possible *Australopithecus/Homo* transitional species, *Homo rudolfensis*, was discovered in 1972 near the shores of Lake Rudolf (now Lake Turkana) in Kenya. Rudolfensis was a contemporary of habilis, having lived approximately two million years ago, but they appear to have been somewhat brainier, with a rounder braincase volume of 775 cubic centimetres and a longer, flat face with less prognathism, all of which were more human-like than Australopithecine. Their molar and premolar teeth were larger than habilis', however, and more Australopithecine than human. Only skull fragments, including facial, jaw, and cranial fragments associated with four separate individuals, have been found to date. No tool artifacts have been found that can be associated with those remains, so nothing is known of rudolfensis' toolmaking capabilities. Paleontologists speculate that because of their relatively large cranial capacity, rudolfensis most likely did fashion tools that were at least as advanced as habilis.

Today, probably because the evolutionary transition was gradual with changes occurring incrementally and, perhaps, even in different subspecies, it remains unclear as to which hominin species marks the end of genus *Australopithecus* and the beginning of genus *Homo*. Likely, there were transitional species or subspecies in addition to habilis and rudolfensis that have yet to be discovered, and each of these subspecies would have displayed different transitional features. It is also likely that these transitional subspecies interbred, and the first *Homo* species may, in fact, have been a hybrid of several transitional subspecies.

The defining anatomical characteristics that distinguish we humans from all previous hominins are a large brain, long legs, manual dexterity, and bipedalism, all of which were "under development" in earlier hominins such as the Australopithecines. Given that many species of hominins had some human-like characteristics at various stages of evolvement, it is often a challenge for paleoanthropologists to draw lines of distinction between species and to identify the first, undisputed member of the genus *Homo*. In truth, it is not so much our anatomy or morphology that defines us, but our culture, which consists of the complex integration and interplay of language, artistic expression, social organization, spiritual beliefs, scientific knowledge, and technology. Since only indirect evidence of most of these cultural elements survive in the fossil record, it is understandable why there remains much uncertainty surrounding the precise evolutionary details of our origins.

CHAPTER THREE

The Trials of the Pleistocene

"Let me embrace thee, sour adversity, for wise men say it is the wisest course."

...William Shakespeare

IN THE LATTER stages of the Australopithecine time on earth, some 2.6 million years ago, the earth was entering a new geological epoch known as the Pleistocene. The first appearance of the *Homo* genus and its subsequent evolution to *Homo sapiens* occurred entirely within the Pleistocene, which ended twelve thousand years ago, about the time *Homo sapiens* tribes first entered North America and completed their occupation of all continents on earth save Antarctica.

The Pleistocene epoch was a time of severe climate change and geographical transformation. Prior to the Pleistocene, the earth had basked in a warm, benign climate that had persisted for more than two hundred million years, but, starting about fifty million years ago, it had been gradually cooling, so at the beginning of the epoch, the climate had reached annual mean global temperatures similar to today's.

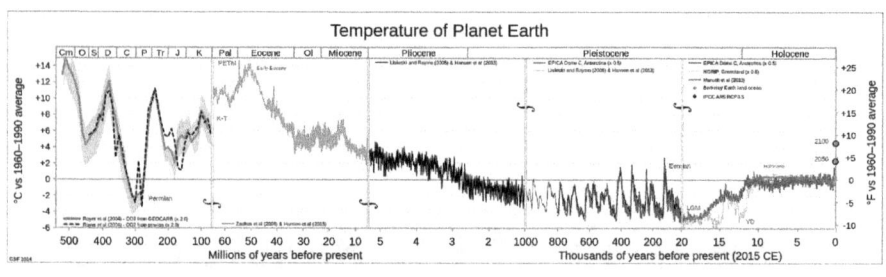

Estimates of the earth's average annual temperature over the last five hundred million years, relative to the 1960–1990 global mean annual value.

Prior to the Pleistocene epoch, the slowness with which the climate changed gave animal species time to evolve and adapt to the resulting changes in environmental conditions. But starting with the Pleistocene, the earth's average annual temperature began to cycle much more rapidly between extreme cold and more temperate periods. During the 2.6 million years of the epoch, there were eleven major and several minor ice ages with an average periodicity of only 130 thousand years, much shorter than anything the earth had experienced throughout the preceding five hundred million years. During that time, flora and fauna had to adapt to extreme climate changes in relatively short periods of time. Many failed to do so and are now extinct.

The ice ages, or *glacials*, were separated by mostly brief, warm periods known as *interglacials*, before the earth was plunged into the deep freeze by the next ice age. This cycle of glacials and interglacials began about 2.5 million years ago and had an average frequency of about one cycle every forty to one hundred thousand years.

During the glacials, global temperatures plummeted, and in the north, they remained well below freezing throughout the year, so atmospheric moisture became trapped on the land surfaces as snow. The snow accumulated and compressed under its own weight, forming vast icefields. At the height of these glacials, the icefields in some places were more than three kilometres deep and covered about 30 percent of the earth's entire land area.

During the interglacials, the glaciers retreated into the high mountain valleys and polar regions, exposing land masses that had previously been under the ice. Temperatures soared, rainfall resumed, sea levels rose, and grasslands, woodlands and rain forests spread northward and southward from the equatorial regions. The northern regions of the earth were transformed from barren icefields into lush, life-nourishing ecosystems, and flora and fauna migrated into the northern regions once again.

These rapidly cycling periods of cold and warmth had a huge impact on the earth's ecosystems and, as a result, on the evolutionary direction of the entire earth's flora and fauna. Such climate change events are ongoing today, and their impact on the earth's ecosystems continues to occur throughout the globe.

During the glacials, evaporated sea water became permanently deposited on land surfaces, and rainfall amounts became much reduced. The sea level dropped, at times by more than one hundred metres below current levels, and millions of square kilometres of coastal shelf became dry land. Neighbouring geographic regions such as Britain and Europe, Indonesia and Malaysia, Siberia and Alaska, the horn of Africa and the Middle East, which are separated by sea water today, were linked by land bridges or island chains during these times.

During glaciation periods, many animal populations in the northern areas were forced to migrate to places where the climate was more hospitable and food resources more readily available. The exposed coastal lowlands and land bridges formed passageways for these migrations, providing access to the more benign environments.

The most recent of the Pleistocene glacials, often referred to as the *Great Ice Age*, began about 125,000 years ago and lasted until approximately 12,500 years ago. At its height, the global average temperature dropped to the lowest levels of the Pleistocene, about 6°C lower than today, and massive icefields covered much of the earth. In North America, advancing glaciers merged, forming the vast Wisconsin glaciation that covered all of Canada and extended over the North Atlantic Ocean and well into the northern United States. In Eurasia, the Weichselian glaciation extended from Scandinavia to northwest Russia.

In the Southern Hemisphere, the Antarctic icefields extended north, covering much of the Southern Ocean. In South America, Andean glaciers merged, forming the Patagonia Ice Sheet. Icefields developed in southern

New Zealand and Tasmania. And in the more tropical regions, ice caps and mountain glaciers developed in places like Western New Guinea, southern Australia, and South Africa.

These massive ice sheets had the effect of shifting major weather systems towards the equator, causing temperatures to fall even at subequatorial latitudes in Africa, south Asia, and Central America. Cool, arid conditions spread across many regions of the globe, tropical rain forests and woodlands shrank, while deserts and grasslands expanded.

Throughout central Eurasia, a vast, frigid, barren steppe land developed. The Mammoth Steppe, as it is called today, extended from northern Spain across central Europe and Siberia into Alaska and the Yukon, creating one of the largest continuous biomes the earth has ever seen.

Since the end of the last Great Ice Age, some twelve thousand years ago, the earth has been in the midst of an interglacial, known as the Holocene epoch, during which the earth's average temperature has remained relatively stable until modern times. This temperature and climate stability were significant factors in the invention and spread of agriculture as the world's first farmers could finally count on regular rainfall and consistent growing conditions from one year to the next.

Since the beginning of the industrial revolution, however, the average global temperature has been rising at an unprecedented rate due to the rapidly increasing amounts of greenhouse gases accumulating in the atmosphere. And recent scientific studies have revealed that human activity is by far the dominant factor in this accelerated warming of the earth. So significant has the human impact been on the earth's climate that some scientists now refer to the present epoch as the *Anthropocene*, an epoch in which human activity is the dominant factor in altering the earth's climate and ecosystems. In the preceding earth's temperature graph, estimates for the near-future values of earth temperature, which take current global warming trends into account, are provided. The estimates were produced by Berkeley Earth, a non-profit organization dedicated to understanding the mechanisms underlying the earth's climate and climate change. If the Berkeley Earth estimates are correct, by 2050, the earth's annual mean temperature will reach levels close to those that occurred during the Eemian interglacial event, the warmest period in the 2.8 million year Pleistocene epoch, when the earth's climate became unusually hot and humid, and tropical flora and fauna extended well into today's more temperate regions. By 2100, the Berkeley Earth models forecast temperatures close to 5°C warmer than today, temperatures the earth has not experienced since the latter part of the Miocene epoch when tropical rain forest extended north of the Arctic Circle.

During the Pleistocene glacial periods, the equatorial regions of the earth acted as safe havens to which both plants and animals retreated as glaciation advanced in the north. These regions provided not only hospitable ecosystems for flora and fauna to live while the north was locked in ice, but they also acted as crucibles of life as new species of flora and fauna evolved there in response to the changing environmental conditions. The most significant of these Pleistocene crucibles was Africa, where early ancestral forms of many animals found throughout the globe today, including humans, first appeared on earth.

The climate change events of the Pleistocene also resulted in large, cyclical changes in the eco-systems of the African continent. During the warm, wet periods, the equatorial rain forests spread into the Sahara and Savanna regions, and when the cooler drier periods returned, the Sahara once again became grassland or desert and the savanna again spread across east Africa. Many lakes along the Great Rift Valley filled during the wet periods and disappeared during the dry. During the harshest of the cyclical environmental conditions, the survival capabilities of early hominins were severely tested, and there were times when life on the savanna became so stressful that they were forced to leave the region altogether. And during the most extreme of these periods, our ancient ancestors came perilously close to extinction.

The Erectines

In contrast to the severe Pleistocene climate events to come, two million years ago East Equatorial Africa would have appeared much like it does today. There were vast areas of savanna, rimmed and interspersed with sparse woodlands and grasslands. In the Rift Valley regions of Tanzania, Kenya, and Ethiopia, there were rivers and lakes, bordered by marsh and denser woodlands. As in modern times, the climate was seasonal when rainy periods would fill lakes and rivers and the savanna and grasslands turned green and lush, alternating with dry seasons when rivers and marshes dried, lake levels dropped or vanished, and the grasslands and savanna withered.

Just as today, vast herds of wildebeest, antelope, zebra, and other ungulates migrated throughout the region in response to seasonal climate changes, preyed upon by packs of hyenas and wild dogs, prides of lions, and families of cheetah and leopard. Herds of elephants and giraffes grazed on the grasslands and woodlands, and rhinoceros, hippopotamuses, and crocodiles could be found around and in the marshes and lakes.

By this time, the Australopithecines had evolved with hybrid anatomical characteristics that were adapted to life in a mixed habitat, and while they were equally at ease foraging in the forest canopy or in the grasslands, they were not adapted to full-time life on the open savanna. With vast herds of wildebeest and antelope, the savanna offered an abundant supply of fresh meat to be sure, but it also presented a harsh and dangerous environment where the ever-present large predators would have been quick to seize upon the opportunity for a relatively easy meal of primate.

The Australopithecines lacked the size, strength, and technology with which to kill and butcher large ungulates or to defend themselves against predators. Consequently, their diet consisted mostly of fruit, seeds, nuts, berries, and only occasionally meat, which would be obtained by killing small mammals or scavenging the carcasses of animals left behind by other animals. Only rarely would they have killed a large animal on their own, and in such cases, the animal would likely have been injured, very young, or otherwise incapable of running away or defending itself.

While the Australopithecines walked upright and were bipedal, their small stature and hybrid anatomy limited their ability to migrate over the long distances required to deal with the seasonal variations in climate that occurred on the savanna. During the cooler and drier seasons, they could only retreat into the safety of the rain forest or woodlands.

Any hominin that was to thrive on the open savanna would have to be bigger, stronger, and faster than the Australopithecines and with the intelligence to outwit other animals. It would have to live in coordinated social groups, and the group would have to communicate and practice cooperative hunting techniques to catch large ungulates and defend themselves against predators. To maintain a diet of predominantly fresh meat, they would need effective weapons to kill large animals and tools to butcher them. Most of these capabilities would require a greater intelligence to master, and it was likely for this reason that natural selection began to favour a bigger, stronger body, and a larger brain with the associated increase in cognition.

Up until the time of the Australopithecines, primate brain mass increased only gradually, and for most primates that increase plateaued about two million years ago at an average of roughly 500 cubic centimetres, about the same size as observed in the larger non-human primates today. Only members of the genus *Homo* continued to experience growth in brain mass after this time period, ultimately reaching the extraordinary volume of the modern human brain, which averages about 1,250 cubic centimetres.

The fact that members of the genus *Homo* are the only primates, and indeed the only animals, to have evolved with an extraordinarily large brain relative to their body mass, continues to intrigue scientists, and there is continuing debate surrounding the question of why that occurred. At first thought, it would seem that an increased intelligence would more or less automatically provide for increased survivability for almost any animal species, but if this were the only factor, the brain size of most animal species would have increased over time.

The answer to this riddle may lie in the fact that the brain is an expensive organ to maintain, consuming more energy than any other organ in the body. In humans, for example, the brain consumes about twenty watts of power, or roughly 20 percent of the body's total energy consumption, and at a rate of ten times faster than the rest of the body on a per gram basis.

For animals with large brains, the disproportionate energy consumption requires a larger caloric intake, and from an evolutionary perspective, the increased caloric requirement could be justified only if the increased intelligence afforded a substantial increase in survival advantages, such as enabling the easier acquisition of food resources, superior defence capabilities, or better strategies for coping with environmental change.

In the case of most primates, which remained in a forested or mixed forest-grassland habitat, food resources were readily available, predators were relatively easily avoided, and coping with climate and environmental change required only migrating slowly with the forests as they advanced or retreated in response to climate change. Hence, for most primates, and indeed most animals, the increase in survivability afforded by a larger brain was not significant enough to justify the increased caloric demands, and their brain size plateaued as a result.

To quote Charles Darwin, "It is not the most intelligent of species that survives. It is the one most adaptable to change."

Early archaic humans, however, chose to spend more time on the open savanna where increased intelligence afforded a distinct survival advantage. Food resources, while plentiful, were more difficult to acquire and predators more difficult to avoid. Survival depended on having advanced group-hunting strategies with effective weapons and good butchering technologies. These characteristics were enabled by a larger brain and its associated cognitive capability.

With the Australopithecine anatomy as a starting point, the evolution of a hominin with the ability to thrive on the open savanna required a rather significant evolutionary leap. Yet, within a relatively short time, perhaps a few hundred thousand years, the selective evolutionary stresses imposed by the savanna environment forged such a hominin. And as the Australopithecines were nearing the end of their time on earth, sometime around 2.0 million years ago, *Homo erectus* appeared on the east African savanna.

Homo erectus, or *upright man*, represented a major step along the evolutionary path toward fully human anatomy. They were tall, about thirty centimetres taller than habilis, and were fully adapted to life on the ground, having elongated legs and shorter arms compared to the size of their torso. They were about the same average height and build as today's average human with males reaching heights of 180 centimetres and weighing as much as seventy kilograms and females maturing slightly smaller.

Compared to the Australopithecines, *Homo erectus*' head featured a more domed cranium, their faces were less protuberant, and they had a slightly more projecting, human-like nose. Their jaws were shorter and more lightly built, resulting in a flatter, shorter face with a pointed chin, and human-like canine and molar teeth.

An artist's rendering of *Homo erectus*.

© 2005 by Encyclopædia Britannica, Inc.

Homo erectus is considered by most paleoanthropologists to be the first true archaic human as their anatomy was very similar to ours, save for head shape and brain size. Their average braincase volume is estimated to have been roughly 800 cubic centimetres when they first appeared, but by the end of their earthly tenure, it had reached about 1,200 cubic centimetres, only slightly less than the average human's today. However, their heads were more elongated and less domed, suggesting that the internal organization of their brains was different.

Homo erectus had lost the long forearms and curved fingers of the Australopithecines and, with that loss, the ability to readily climb trees and forage in the forest canopy. But with longer legs and other anatomical changes to the shoulders, rib cage, and waist, their new body shape offered improved balance and made it possible to walk over longer distances without tiring.

Perhaps, more importantly, *Homo erectus* could run. While compared to most other animals, predator or prey, on the savanna, *Homo erectus* was not a fast runner, they could run continuously for much longer periods of time than any of the other animals. And evidence shows that they quickly learned to use this ability to their advantage as meat became the dominant component of their diet.

On the African savanna, the primary focus of *Homo erectus*' hunting and foraging would have been the abundant supply of meat in the form of the vast herds of ungulates—wildebeest, antelope, and zebra, to name only a few. A developing taste for the meat of these larger animals motivated *Homo erectus* bands to develop new techniques for hunting and better technology for killing and butchering them. In this regard, scientists speculate that *Homo erectus* likely learned to practice persistence hunting as a strategy to prey upon large ungulates on the open savanna.

Persistence hunting is a technique in which hunters, who can run for long periods of time but are slower than their prey over short distances, use a combination of running, walking, and tracking to pursue prey over long periods of time. On the savanna, most ungulates are much faster than humans over short distances but are incapable of maintaining such speeds over long distances and must stop to catch their breath at frequent intervals, particularly in the hot African sun. Hence, after a few hours or even days of persistent pursuit, during which the hunters prevent the pursued animal from stopping to rest, the animal eventually succumbs to exhaustion, at which point the hunters can close in for a kill.

In Africa, as *Homo erectus* bands spent more and more time running across the savanna in the pursuit of prey, evolutionary stresses favoured hairlessness, which allowed for better cooling in the hot African sun. And over successive generations, the African *Homo erectus*, referred to by most paleoanthropologists as *Homo ergaster*, lost most of their fur. Genetic studies that compare the genes associated with hair and skin tone in chimpanzees with those of humans today indicate that *Homo ergaster* probably lost their fur sometime around 1.2 million years ago.

While hair loss would have been beneficial to *Homo ergaster* in terms of thermoregulation, it created another dangerous vulnerability, skin cancer. The intense tropical African sun is dangerous to exposed skin, especially fair skin, leading to a variety of skin cancers, the deadliest of which is melanoma. As *Homo ergaster*'s fur thinned, the stresses of natural selection darkened their skin with a protective layer of melanin. By about one million years ago, African *Homo ergaster* had been evolutionarily transformed from a hairy, fair-skinned primate into a naked, dark-skinned, nearly modern-looking *proto-human*.

But while it is an important factor, a bare skin in itself does not ensure efficient thermoregulation of the body, and hairlessness was only one of the important physiological changes that *Homo ergaster* needed to become the world's supreme long-distance runner.

More importantly, they needed a means to thermoregulate their bodies to prevent dangerous overheating of the brain, muscles, and other internal organs as they ran across the savanna in extreme heat. In response, the evolutionary process equipped *Homo ergaster* with an extremely efficient, whole-body sweating mechanism that is unique in the animal kingdom. Sweating cools the body through the production of liquid on the skin's surface that then evaporates, drawing heat energy away from the skin and cooling the interior of the body.

Homo ergaster and we humans today are unique in the mammalian world in that we possess an extraordinarily large number of eccrine glands, between 2 and 5 million of them, distributed all over the body, which can produce about a litre of thin, watery sweat per hour during vigorous exercise. They are located relatively close to the surface of the skin and discharge sweat through pores which then evaporates to eliminate excess heat very efficiently.

The superior thermoregulatory capabilities with which *Homo ergaster* evolved, hairlessness and the ability to sweat, enabled them to run long distances without succumbing to heat exhaustion, and as a result, their hunting efficiency improved and, over time, their diet became predominantly meat.

When bands of *Homo erectus* first appeared on the savanna, they came equipped with the Oldowan tool set developed by their predecessor, *Homo habilis*. But as hunting success improved, a need arose for better tools to butcher the much larger and tougher animals that *Homo erectus* could now kill, and the paleontological record shows that such improvements were under development as early as 1.8 million years ago. Shortly after *Homo erectus'* arrival, more sophisticated stone implements with sharper and straighter edges began to appear, and by 1.5 million years ago, their tool kits were sufficiently advanced to be considered a new toolmaking tradition, referred to as Acheulean, named after the types of tools found at the Saint-Acheul quarter of Amiens in northern France.

The most common Acheulean tools were biface hand axes, made from either rock cores similar to those used in the Oldowan tradition but with flaking on both sides or from large, knapped flakes that were rounded on one end of the flake to make a handle. Both techniques produced a sharper, more uniform blade, giving the tool a symmetrical, teardrop shape. Such tools were likely multipurpose implements that were used for chopping or carving wood, digging up roots and bulbs, butchering animals, and cracking nuts and bones.

A hand axe of the Acheulean tradition.

© *José-Manuel Benito Álvarez*

While hand axes are the most diagnostic of Acheulean tools, *Homo erectus* also manufactured a relatively wide variety of stone tools that included choppers, cleavers, and hammers as well as flake knives and scrapers. Furthermore, their toolmaking expertise progressively improved over the course of their term on earth, and towards the end, they were skilled and extremely prolific craftsmen. Habitation sites littered with tens of thousands of discarded stone tools, dating to approximately five hundred thousand years ago, have been found throughout most of *Homo erectus'* geographic range in Africa and Eurasia.

Homo erectus' diet also evolved over time. Initially, like their Australopithecine ancestors, their diet would have consisted primarily of plant parts, tubers, small animals, and occasionally scavenged meat and eggs. But after a time, as brain size and intelligence grew, their hunting prowess improved, and they began to hunt larger prey. Based on food-related artifacts found at sites throughout Eurasia, by about seven hundred thousand years ago, *Homo erectus'* diet included both small prey, such as birds, turtles, rabbits, rodents, fish, and mollusks, and larger animals such as wild boar, sheep, rhinoceros, buffalo, and deer. It appears that by this time, *Homo erectus* had a broad omnivorous diet that included virtually every edible animal and plant in whatever environment they found themselves and had developed the weapons and tools required to support it.

It remains unclear exactly where *Homo erectus* came from, evolutionarily speaking, as no remains have been discovered that clearly represent a transitional species between the Australopithecines and Erectines. The erectine anatomy represented a substantial change from that of the Australopithecines, and the changes appear to have evolved in the relatively short time frame of only a few hundred thousand years. But based mostly on the observation that the later Australopithecines were already showing signs of an increasing brain size, most paleoanthropologists believe that *Homo erectus* evolved in Africa and descended from one of *Australopithecus sediba, habilis,* or *rudolfensis* with habilis having the most support. This theory is somewhat problematic, however, as habilis was a contemporary of erectus for most of the former's duration on earth, and apart from increased brain size, habilis, rudolfensis, and sediba were all anatomically much closer to the Australopithecines than to *Homo erectus*.

Another enigmatic aspect of *Homo erectus'* origins is that various erectine subspecies seem to have appeared more or less simultaneously, in widely separated geographical regions throughout Africa and Eurasia. So rapid was their apparent dispersal that several conflicting theories have been put forward to explain how the species could have appeared in such widely separated regions in so short a period of time.

Some paleoanthropologists cite climate and environmental change as the reason why the species was forced to migrate quickly to more hospitable places in order to survive. Others speculate that the species may have appeared in Africa somewhat earlier than the fossil record indicates, perhaps as early as 2.2 to 2.3 million years ago, and that bands of these very early erectines began to disperse throughout Africa and Eurasia even earlier than two million years ago.

Still others have suggested that the erectine origins may have been in central or southeast Asia or that they may have evolved independently at multiple locations throughout Africa and Eurasia at about the same time. Supporters of these alternate origin theories cite the possibility that an Australopithecine migrated out of Africa about three million years ago and subsequently evolved into an erectine at multiple locations throughout Eurasia. No clear fossil evidence has been found to support any of these theories, however, and most scientists still believe that *Homo erectus* first evolved in Africa and subsequently migrated into Eurasia.

There is also continued debate on whether the various erectine discoveries throughout Africa and Eurasia are even members of the same species. Paleoanthropologists use the standard taxonomical classification system of biology to determine the evolutionary relationships of fossilized hominin remains. Truth be told, however, this taxonomical system is an artificial construct, invented to provide an ordered system for the identification and comparison of species, and the evolutionary process adheres to no such distinct classification system. New species do not suddenly appear in *quantum leaps*, but rather they continuously evolve through a series of incremental changes, some significant, but most relatively minor. New species emerge only when the accumulated changes are substantial enough as to preclude successful interbreeding.

Furthermore, it must be acknowledged that only the tiniest fraction of the remains of once living creatures end up as fossils, and of those, no doubt only a very small percentage are discovered. There were likely hundreds of millions of hominid individuals that lived between the time of Proconsul and the appearance of Homo sapiens, representing a continuum of species of which we have only a few hundred snapshots from which to reconstruct our evolutionary tree. Hence, it is often a challenge for scientists to determine whether a given set of remains, which in most instances consists of only a few bone fragments, represents a separate species or just a slightly different individual member of a known species. This problem is particularly acute in the case of *Homo erectus* because of their wide geographical distribution and the extraordinarily long time span of their existence. It is no wonder, therefore, that there is still some disagreement amongst paleoanthropologists on the classification of the various erectine remains found throughout Eurasia and Africa.

Perhaps in time, future paleontological discoveries will clarify *Homo erectus'* true geographical and evolutionary origins, but for now, most evidence points to the likelihood that the species originated in the Great Rift Valley of east Africa, sometime around 2.0 million years ago.

Over the last one hundred years or so, there have been many discoveries of erectine remains, which at the time of their discovery, were typically classified as separate *Homo* species. *Homo ergaster* is the taxonomical name given to most erectine discoveries on the African continent; *Homo antecessor* to early discoveries in Spain, France, and England; *Homo georgicus* to specimens found in Dmanisi, Georgia; and *Homo lantiensis* to specimens found at Lantian, China. Popular local names given to some of these specimens are Turkana boy, Gran Dolina boy, Mojokerto child, and Yuanmou man.

A comparative study of these remains in recent times, however, has convinced most paleoanthropologists that they are subspecies of *Homo erectus*, as they all exhibit only minor variations from one another, and those differences lie within the range typically observed within a single species.

Regardless of where and when they first appeared, the erectines were enormously successful from an evolutionary perspective, having lived from about two million to one hundred fifty thousand years ago. They lived entirely within the Pleistocene and proved remarkably resourceful at dealing with the massive climate change events that characterized the epoch. Given their longer legs and advanced bipedal locomotion, the erectines were highly mobile, and immediately upon their first appearance, it seems bands of *Homo erectus* began to migrate extensively throughout the African continent and beyond. Such migrations were most likely in response to changing environmental conditions, and enhanced mobility was, no doubt, a major reason why the species survived the ofttimes harsh and highly fluctuating climate of the Pleistocene epoch.

By approximately 1.8 million years ago, *Homo erectus* could be found living throughout the Great Rift Valley of east Africa, from the Awash Valley region of Ethiopia to the limestone caves of South Africa. And sometime between 2 and 1.8 million years ago, bands of *Homo erectus* wandered northeast across the Levantine corridor into the Middle East, marking the first time a hominin is known to have permanently left the African continent and survived. From the Middle East, *Homo erectus* bands fanned out into Eurasia and by roughly 1.5 million years ago, they had become widely dispersed across Europe and central and southeast Asia, from Atapuerca in northern Spain to the Sangiran caves on the island of Java, Indonesia. A summary of the important sites where *Homo erectus'* remains have been found is shown in the following Table One. The table illustrates how widely *Homo erectus* became dispersed in a relatively short period of time and how their brain size slowly increased over time.

LOCATION	TAXONOMICAL DESIGNATION	LOCAL NAME	ESTIMATED AGE (KYA)*	CRANIUM VOLUME (CC)**
Africa				
Olduvai Gorge, Tanzania	Homo ergaster		1900 - 700	1067
Swartkrans, South Africa	Homo ergaster		1800 - 1500	600 to 650
Koobi Fora, (Turkana) Kenya	Homo ergaster	Turkana Boy	1600 - 1500	880 - 910
Bouri, Awash Valley, Ethiopia	Homo ergaster		1000	800
Drimolen Cave, South Africa	Homo ergaster		2040 - 1950	
Europe/Central Asia/Middle East				
Dmanisi, Georgia	Homo erectus		1800 -1700	
Ubeidiya, Israel	Homo erectus		1600 - 1400	
Spain, France, England	Homo antecessor	Gran Dolina Boy (Spain)	1200 - 800	1000 - 1150
Southeast Asia				
Sangiran, Indonesia	Homo erectus		1800 -1600	813 - 1059
Modjokerto, Indonesia	Homo erectus	Mojokerto child	1500	
Trinil, Indonesia	Homo erectus javanensis	Java man	1500 - 700	900
China				
Yuanmou, China	Homo erectus	Yuanmou Man	1700	
Lantian, China	Homo erectus lantianensis	Lantian man	1600	780

* - thousands of years ago
** - Cubic centimeters

Table One - Geographic dispersal of early *Homo erectus*

The first discovery of erectine remains was made in 1891 by Eugene Dubois on the banks of the Solo River in Indonesia. Dr. Dubois and his team uncovered a tooth, skullcap, and thigh bone and published their findings as "the missing link" between apes and humans. While today's paleoanthropological community has pretty much abandoned the idea of a single link species, *Homo erectus javanensis*, or Java Man, is considered the type specimen for *Homo erectus,* even though the Java Man remains have been dated to only one million years ago.

A remarkable early *Homo erectus* specimen was discovered by paleontologists Kamoya Kimeu and Richard Leakey near the eastern shore of Lake Turkana, Kenya, in 1984. The specimen, named Turkana Boy, has been dated to between 1.6 and 1.5 million years old and is that of a young boy, estimated to be between eight and ten years old at the time of his death.

The Turkana Boy specimen is composed of 108 bones, making it one of the more complete erectine skeletons discovered to date. Turkana Boy stood about 160 centimetres tall, and it is estimated that, in adulthood, he would have reached the somewhat remarkable height of 185 centimetres and weighed 68 kilograms.

Because they were so widely separated geographically, and their overall population was small, estimated by paleoanthropologists to number in the few tens of thousands globally, bands of early *Homo erectus* became isolated from one another for long periods of time. They lived in very diverse environments, ranging from the tropical rain forest of southeast Asia; to the hot, dry savanna regions of Africa; the cold, temperate forests of Europe; the barren steppes of central Asia; and the forests and woodlands of central China. Under such conditions, variations in skin tone, hair and eye colour, facial features, and anatomical proportions would most likely have evolved, and some paleoanthropologists speculate that these variations are reflected in the different racial characteristics observed in people throughout Eurasia and Africa today.

In spite of these variations, however, *Homo erectus*' general anatomy retained the defining characteristics of a single species, and bands of *Homo erectus* individuals would have interacted socially as members of the same species whenever they met, just as human individuals from different races do today.

The Discovery of Fire

"The best way to make a fire with two sticks is to make sure that one of them is a match."

...Will Rogers

At the base of a brush-covered hill in South Africa's Northern Cape province, a massive stone outcropping marks the entrance to one of humanity's oldest known dwelling places, Wonderwerk Cave. Based on the remains and artifacts it contains, it appears that the cave was home to erectine bands for almost two million years. And it was in Wonderwerk Cave that the earliest, definitive evidence of the controlled use of fire was found in the form of a fire pit, around which were charred bones and plant ash remains, radiocarbon dated to approximately one million years ago.

While evidence of charred material has been found in cave sites dating as far back as two million years, this older evidence is controversial as it cannot be determined whether the charred remains had been deliberately cooked, were carried into the site after being charred in a wildfire, or were possibly even deposited in the cave by natural processes. In Wonderwerk Cave, however, the evidence for the controlled use of fire by *Homo ergaster*, about one million years ago, is much stronger and has wide scholarly support.

It cannot be determined how *Homo ergaster* made fire one million years ago, however. It is possible that they had learned how to make it from scratch by this time, or they may have acquired fire opportunistically from a naturally occurring source, bringing it back to the cave where it could have been kept burning for several days or even weeks at a time.

One can only speculate as to how humans would have first discovered the benefits of fire, but it likely happened opportunistically with the fortuitous consumption of the charred remains of animals that had been burned in a wildfire. It is easy to imagine *Homo erectus* bands wandering through a landscape after a wildfire had run its course and finding the partially charred remains of freshly killed animals, tasting them out of hunger or curiosity, and discovering that cooked flesh was rather agreeable—tastier, easier to chew, and more digestible than raw meat. It wouldn't be long before these same early humans would habitually search for naturally cooked animal remains following wildfires and would bring them back to their campsite for communal consumption.

It would likely not have been long after that before some bright individual got the idea of bringing still-glowing embers from a recent wildfire back to the communal cave to start a *controlled* fire in a firepit. The fire could then be kept burning by regularly plying it with wood and dry grass, perhaps for days or even weeks. Thus, the benefits of fire for cooking meat and for providing heat and light could have been discovered.

Exactly when erectines learned to make fire from scratch and how they did it cannot be known with any degree of certainty. But ample evidence exists at habitation sites throughout the African continent and in the

Middle East that they had learned how to make fire on demand at least by seven hundred thousand years ago and, as the evidence at Wonderwerk suggests, perhaps well before that.

Starting a fire from naturally occurring materials is an acquired skill, as anyone who has tried it can testify. Learning how to make fire would have been a major intellectual challenge for early humans who had no knowledge of what materials to use or even that such a thing was possible. It is no surprise that it took hundreds of thousands of years for our ancestors to transition from the first opportunistic use of fire to making fire on demand.

Fire can be kindled in a variety of ways from natural materials such as wood and stone, but all methods require ingenuity, patience, and practice to discover and perfect. A common method used today, apart from using a match, is to strike a piece of iron with a hard stone such as flint, producing hot chips of iron or "sparks," which can be directed into tinder to start a fire. The palaeolithic equivalent of this would have been to use flint and a naturally occurring iron-based rock such as pyrite or marcasite. But while marcasite has been found not far from Wonderwerk Cave, no evidence of its use for making fire has been found at any paleontological site to date.

Another method, known as the *Fire Plow*, uses a wooden stick cut to a dull point and a long piece of wood with a grooved indentation down its length. The point of the first piece is rubbed with force against the groove of the second in a rapid plowing motion, producing wood dust in the process. The friction between the two pieces of wood heats the wood dust until it forms a hot, smouldering coal, which is then directed into tinder to ignite it.

Cooking was undoubtedly the first and most beneficial consequence of mastering fire. In addition to improving flavour and aiding digestion of familiar foods, cooking expanded *Homo sapiens'* diet by softening otherwise inedible foods, such as hard nuts and tough, fibrous, root vegetables. Cooking meat destroyed harmful parasites and pathogens and broke down some larger proteins, rendering them more available for quick uptake and use by the body, so it could build its own proteins. Cooking removed some toxicities, improved flavour, and eased digestion, while at the same time facilitated the body's absorption of nutrients and energy. Eating cooked meat produces more energy, particularly beneficial for our oversized brains, and in this regard, some scientists theorize that cooking may even have been a contributing factor in the evolution of *Homo sapiens*' enlarged brain.

Of course, most of these benefits derived from cooking food were unbeknownst to early humans. They simply ate what tasted good and was easy to chew and digest. But at times when food was scarce, cooking enabled them to eat a wider range of foods and extract more energy and nutrition from the food they did manage to secure. Such benefits enabled them to endure greater hardship and to survive in more difficult circumstances brought on by severe climate events or when they were forced to migrate into new ecosystems with unfamiliar flora and fauna.

The mastering of fire produced other benefits that would have been quickly discovered as well. Fire provided a source of warmth and light, enabling human industry and socializing to be conducted into the darker and cooler hours of the night. It also provided a measure of security as most predators had an instinctive fear of fire and would have avoided entering caves or campsites where a fire was burning.

There is evidence that erectines used fire to harden the tips of wooden spears as early as four hundred thousand years ago. The process involved making a point on a wooden pole by rubbing it against a smooth stone or whittling it with a flint blade. Next, the point was heated in a fire to bring the moist pitch in the wood to the surface, producing a light coating of the charred wood and pitch mixture on the surface. The tip was then polished with a soft stone so that fine stone particles were added to the pitch and charred wood. Subsequent firings and polishings eventually formed a hardened glaze on the tip that is comparable in hardness to a metal point.

In essence, the erectines used fire for the same things for which we use energy today: for cooking, lighting, central heating, tool fabrication, and security. The introduction of controlled fire into daily routine was a major milestone in human evolution and one of the most important foundational technologies upon which human culture was enabled.

Today, we have learned to harness and produce *fire* in the form of energy in many different ways, but our dependence on and attachment to fire remains essentially unchanged from prehistoric times. We still rely, for the most part, on coal, gas, wood, or oil-fired furnaces to heat our homes, and some still jokingly refer to "stoking up the fire" when they turn up the thermostat. We talk about the firepower of our weapons and the candlepower of our lighting and who doesn't still enjoy having a crackling fire nearby.

It seems there is a primeval appeal about an open fire, whether it's a burning log in our fireplace or a campfire in the wilderness, that both fascinates and comforts us, and that is quite unrelated to its more utilitarian benefits. And as that group of erectines sat around their fire in Wonderwerk Cave a million years ago, it is quite possible that they felt much the same way.

The Next Generation of Erectines

As *Homo erectus* continued to evolve throughout Africa and Eurasia, their cranium size steadily increased, so between 750 and 250 thousand years ago, a *second generation* of erectines was emerging with an average braincase volume approaching that of a typical modern human.

With increased intelligence came more advanced technology and a more effective social organization, both of which enhanced survivability through the practice of more effective hunting and defence strategies. As a result, the population of *Homo erectus* expanded worldwide as evidenced by the relatively large increase in the number of *Homo erectus* sites that have been discovered throughout Africa and Eurasia dated to that period.

By this time, erectine toolmaking skills had advanced from the Oldowan to the Acheulean tradition, and they had mastered the use of fire. Such technological advances were not limited to Africa as the tools discovered with *Homo erectus* remains dating to this period show similar advances in most habitats throughout Eurasia. There is also evidence of the controlled use of fire at places as distant as Suffolk, England, and Zhoukoudian, China.

At the time that bands of *Homo erectus* began dispersing throughout Eurasia, they were very likely still fur-covered with fair skin, much like today's chimpanzees, since genetic analysis has shown that African *Homo erectus*, or *ergaster*, didn't evolve to hairlessness with a dark skin until approximately 1.2 million years ago. Because a fur coat and fair skin offer a benefit in a northern climate where sun was less intense and temperature lower, it's likely that the Eurasian *Homo erectus* maintained those characteristics. In more temperate environments, the ultraviolet component of sunlight is an important factor in manufacturing vitamin D in the body, and the increased sun exposure afforded by fair skin would have been a survival advantage.

Hundreds of thousands of years later, when the black-skinned, hairless descendants of *Homo ergaster* left Africa for the more northerly regions of Eurasia, their skin tone lightened once again to allow for better absorption of vitamin D. But since our ancient ancestors were making fur clothing by that time, the survival advantage afforded by a natural fur coat was diminished, and they remained largely hairless into modern times.

Many second-generation erectine discoveries have been made throughout Eurasia and Africa, indicating that by about five hundred thousand years ago, populations of erectines had increased substantially. Table Two provides a summary of important discoveries dating to this time and how brain size had increased to nearly modern human size.

LOCATION	TAXONOMICAL DESIGNATION	LOCAL NAME	AGE (KYA)*	CRANIUM VOLUME (CC)**
Africa				
Bodo d'Ar Ethiopia	Homo erectus heidelbergensis		800-600	1200 - 1325
Ternifine, Algeria	Homo erectus heidelbergensis		700	
Broken Hill, Zambia (Rhodesia)	Homo erectus rhodesiensis	Rhodesian Man	800 - 125	1100
Europe/Middle East				
Mauer, Heidelberg, Germany	Homo erectus heidelbergensis		500	
Arago Cave, Tautavel, France	Homo erectus heidelbergensis	Tautavel Man	450	1150
Happisburgh, Norfolk England	Homo erectus heidelbergensis		780	
Petralona Cave, Greece	Homo erectus heidelbergensis		400 - 250	1230
Boxgrove, England	Homo erectus heidelbergensis	Boxgrove People	524 - 478	
Ceprano, Italy	Homo erectus heidelbergensis		900 to 690	
Atapuerca, Spain	Homo erectus heidelbergensis		800 - 300	1125 - 1390
Steinheim an der Murr, Germany	Homo erectus heidelbergensis	Homo steinheimensis	400 - 300	1100
Swanscombe, Kent, England	Homo erectus heidelbergensis	Swanscombe Man	400	1325(?)
Vértesszölös, Hungary	Homo erectus heidelbergensis	Samu	475 - 250	1400(?)
Jaljulia, Israel	Homo erectus heidelbergensis		500	
Southeast Asia				
Ngandong, Indonesia	Homo erectus soloensis	Solo Man	550 - 143	1,013 to 1,251
China				
Zhoukoudian, China	Homo erectus pekinensis	Peking Man	770 - 300	1030 to 1140
* - thousands of years ago				
** - cubic centimeters				

Table Two - Geographical dispersal of the second-generation erectines

Homo erectus heidelbergensis

Of all the second-generation erectines, the most ubiquitous appears to have been *Homo erectus heidelbergensis*, which lived throughout Africa and western Europe from about seven hundred to three hundred thousand years ago. Heidelbergensis, so named because the first discovery of its remains was made at Mauer, near Heidelberg, Germany, shared most features with early *Homo erectus* but with a few variations trending further towards modern human anatomy. Heidelbergensis had more modern, human-like dental features and a larger cranium with a brain size equal to about 90 percent of that of today's average human. They also appear to have been well adapted to life in colder climates as those living in the more temperate regions were more heavily built with a shorter, wider body that would have retained body heat more efficiently than their African cousins.

Heidelbergensis was a prolific toolmaker. They had inherited the Acheulean tradition of tool fabrication, and evidence shows a gradual refinement in the quality and craftsmanship of their tools over the course of their time on earth. There is evidence that, in addition to hand axes, Heidelbergensis was using fine flake tools as points and scrapers. By about five hundred thousand years ago, they had learned to haft a stone flake onto the end of a wooden shaft, making the world's first stone-tipped spear with which they could kill larger animals.

Evidence found at multiple archaeological sites in Europe, such as Torralba and Ambrona in Spain and St. Esteve-Janson in France, indicate that Heidelbergensis lived in small social groups, typically in caves or rock shelters, and had mastered fire-making. At the Beeches Pit site, in Suffolk, England, uranium series and thermoluminescence dating place the use of fire to about 415 thousand years ago.

There is much debate surrounding the cognitive and language capabilities of *Homo erectus heidelbergensis*. Analysis of the inside of an intact Heidelbergensis skull has shown that the area of the brain associated with speech, known as the Broca's Area, was enlarged, suggesting a degree of linguistic capability. How sophisticated their vocabulary would have been or whether they had any form of complex grammatical language is unknown. However, some paleoanthropologists argue that as Heidelbergensis' toolmaking technology became more advanced, some form of grammatical language would have been required to teach others the subtleties of the new knapping techniques.

The existence of language, in turn, implies a degree of abstract thinking that also underlies the development of artistic expression. There is very little evidence of any form of artistic expression displayed by Heidelbergensis, but at the archaeological site at Twin Rivers, a complex of caves in southern Zambia, large quantities of ochre in the form of pieces of hematite, limonite, and specularite have been found, ranging in age between 400 and 266 thousand years old, the time period during which *Homo heidelbergensis* lived there. Such minerals would have been of no use other than for the production of pigments, which can be made by grinding the material into a fine powder and mixing it with water or rendered animal fat. A good range of coloured pigments, including red, yellow, brown, and purple, can be made in this way. The source of the ochres found in the Twin River caves was some distance away, suggesting that the ochre would have been deliberately carried to the cave site. Archaeologists speculate that the pigments were used for some form of ritualistic body painting or possibly cave painting, although no evidence of such use has survived from this period.

And then there is the Berekhat Ram Pebble, a small lump of volcanic lava that vaguely resembles a human female figure. The pebble was discovered on the Golan Heights in the Levant and has been dated to between 280 and 250 thousand years ago. The pebble measures thirty-five millimetres long and has had at least three grooves incised on it by a sharp-edged stone. One deep groove encircles the narrower, more rounded end of the pebble, demarcating the head. Two shallower, curved grooves that run down the sides demarcate the arms. Understandably, the figure is highly controversial, partly because of the vagueness of its resemblance to the female figure but more so because of the significance it represents if it was deliberately modified by Heidelbergensis.

The *Venus of Berekhat Ram.*

Some argue that it was simply a naturally occurring stone and the striations could have been made by any number or natural processes. Others, however, have examined the pebble under a microscope and have concluded that the striations were indeed human made. If this is true, the *Venus of Berekhat Ram* represents the oldest example of figurative art ever found.

The Berekhat Ram Pebble is the only sample of scoria stone found at the site, so it must have been carried to the caves from elsewhere. Scoria is not suitable for toolmaking, so one must wonder why it would have been kept. Even if the striations are made by natural processes, it appears the stone was kept for some non-functional use, and the most likely reason would have been curiosity about its resemblance to the female form, which in itself, is indicative of relatively advanced abstract thought.

The Boxgrove People

Five hundred thousand years ago, prior to the formation of the English Channel, the British Isles constituted a large peninsula extending northwest out of continental Europe into the Atlantic Ocean. The Seine and Thames Rivers flowed into a huge glacial lake in what is now the North Sea, and a river, known as Fleuve Manche, flowed westward from the lake. The Fleuve Manche wound its way southwest through chalk hills until it emptied into the Atlantic Ocean through an estuary located near present-day Chichester, West Sussex, England.

During the Pleistocene interglacials, when the climate was warm and wet, the river swelled into a giant super river and began to cut a massive trench through the soft, soluble chalk along today's Strait of Dover to the Atlantic Ocean. The white cliffs of Dover mark the remnant banks of this once great river.

During the Pleistocene glacials, icefields covered much of the peninsula, and glacial movement further eroded the channel that had been carved out by the river. The glacials waxed and waned for almost five hundred thousand years until, about nine thousand years ago, they finally retreated northward into Scandinavia, exposing the British Isles as they appear today with the Thames River flowing into the North Sea to the east, and to the south, the Seine flowed into the newly formed English Channel.

Prior to the formation of the English Channel, animals, including archaic humans, moved freely between northern Europe and Great Britain, wandering northward during the warm Pleistocene interglacials and retreating southward during periods of glaciation. During the glacials, Britain was void of all but the hardiest of flora and fauna, but during the warmer periods, the area became a mixed habitat of forests and wetlands where a large variety of animals was abundant.

Five hundred thousand years ago, Europe was in the midst of such an interglacial.

About five kilometres northeast of Chichester in present-day England lies the village of Boxgrove, a small civil parish set just inland from the south coast of Britain. The village has a small population of just under one thousand people, but it boasts a long history of continuous settlement. Ruins of a Benedictine monastery founded in 1115 are still present, and a thirteenth-century parish church remains in regular use today.

Boxgrove village is indeed old, but these medieval structures belie the true age of human settlement in the area. Five hundred thousand years ago, at nearby Eartham quarry, a band of *Homo heidelbergensis* individuals, known by paleoanthropologists today as the Boxgrove people, lived at the base of a massive, one-hundred-metre-high chalk cliff. From their dwelling at the base of the cliff, the Boxgrove people looked across a broad coastal plain that was home to a bounty of large animals, including red deer, bison, elephants, rhinoceros, and horses, along with the ever-present predators—wolves, hyenas, and lions. Supporting this bountiful wildlife, the plain was dotted with freshwater ponds, and beyond it lay the estuary of the Fleuve Manche, a wetland of salt marshes and grasslands where waterfowl, including the now-extinct great auk, foraged along the shoreline.

While nothing remains of dwellings or firepits at the Boxgrove site, it is likely that, given the variability of the northern climate in which they lived, the Boxgrove band would have had shelter of some kind. Perhaps they had found a natural cave at the base of the chalk cliff, or they may have constructed some kind of shelter of wood and animal hides. However, the chalk cliff has long since eroded, and wooden structures would have similarly disappeared over time. It is likely also that they had mastered the use of fire, if only opportunistically, for cooking their food and keeping warm during the cooler northern evenings. In any event, it appears the land was bountiful, and the people lived well, feasting mostly on horse and rhinoceros meat and enjoying nuts and berries as the seasons provided.

After a time, the water levels in the Fleuve Manche began to rise, likely due to a combination of increased glacial melt upstream and perhaps heavier than usual rains. The river overflowed its banks and flooded the valley right up to the base of the chalk cliffs, and the people were forced to vacate the area, leaving many of their possessions, such as they were, behind. As the water rose and engulfed the site, any wood and thatch shelters, along with charred wood from firepits, were washed away, but the heavier artifacts such as flint tools and butchered animal bones remained in place. As the rising water covered them, a thick layer of chalky silt was laid down, preserving them just as they were left some five hundred thousand years ago. Within a fairly short time, the entire locale where the band lived, worked, and hunted was covered in silt, preserving a rare archaeological site that contained human artifacts at multiple activity sites, providing remarkable insight into how the Boxgrove people lived five hundred thousand years ago.

By modern times, the high chalk cliffs had been eroded away to just a few metres in height, and the silt-covered site had, in turn, been covered by several metres of gravel and glacial debris. But as luck would have it, the gravel became useful in the twentieth century for building new roads, and the gravel layer was excavated and hauled away, exposing the layer of silt that was of no commercial use. The silt remained largely undisturbed, until archaeologists Mark Roberts and Michael Pitts began to excavate the site in the 1980s.

The Eartham Quarry archaeological site at Boxgrove has proven to be a rich trove of five-hundred-thousand-year-old artifacts, including over three hundred Acheulean hand axes and the butchered remains of several animals, providing unusually detailed insight into the behaviour and daily routine of early *Homo heidelbergensis*.

Archaeological excavations have yielded multiple sites where the Boxgrove people conducted routine activities in an organized fashion. One such site was a flint quarry located at the base of a nearby cliff where chunks of flint had eroded out of the chalk wall and fallen to the ground below. Quantities of flint cobbles lie at the cliff base, and both finished and partially finished hand axes have been found, along with the discarded debitage left over from their production. Some of the cobbles were partially flaked and discarded as if they had been rejected for poor quality, while other axes, presumably completed, had been removed for use elsewhere.

The exact places where people knapped their hand axes have been preserved, with flakes of flint still lying where they fell. In some instances, the arrangement of the debitage that accumulated during the knapping process allowed researchers to discern the position of the toolmaker's legs as he or she went about their work. It appears also that the Boxgrove people may have learned to fashion more than just hand axes from flint cobbles. A small pile of larger flakes has been found that appears to have been separated from the other debitage piles for use as scrapers, knives, and possibly spear tips.

Another site, located at the edge of what had been a freshwater pond, must have been a favourite animal butchering place. A large number of hand axes and smaller tools have been found there along with the butchered remains of red deer, rhinoceros, bear, giant deer, and horse. The freshwater pond would have attracted animals of all sizes and would therefore have been an excellent place for Boxgrove hunters to lie in wait for the animals, killing and butchering them on-site.

Very little by way of human remains have been uncovered at Boxgrove, suggesting that the people that lived there simply evacuated the area rather than being overcome by a sudden disaster. However, a massive tibia and two incisor teeth were found and based on the size of the tibia, the individual, deemed to have been a male, is estimated to have weighed over ninety kilograms. He has been given the nickname Roger, after Roger Pedersen, the volunteer worker who first spotted the fossil. Prehistoric Roger is estimated to have stood about 1.8 metres tall and was approximately forty years old at the time of his death. Life, at least for Roger, must have been good at Boxgrove five hundred thousand years ago.

A few hundred metres away from the pond, a third site has yielded the butchered remains of a large horse, surrounded by piles of debitage, which suggests that tools were manufactured on the spot, presumably to butcher the horse where it was slain. Since no complete tools have been found at the site, they must have been carried off, and the site was a one-off butchering place. Perhaps the horse was too heavy to carry to their normal butchering spot at the pond, even for Roger, so it was butchered where it was slain.

From the discoveries at Boxgrove, a picture emerges of a primitive human society that lived in relative peace with the bountiful flora and fauna of a coastal plain satisfying all of its needs. The individuals manufactured fine tools from the flint found at the base of a nearby cliff and cooperated in the hunting and butchering of a variety of animals that came to drink at a nearby watering hole. While the physical evidence has long since been washed away, they likely lived in a cave shelter and cooked their food over a firepit. From the vantage point of the campsite, located at the base of a massive chalk cliff, they could socialize and entertain each other while looking out over the Fleuve Manche delta. Perhaps they enjoyed watching the sun as it slowly set on the distant horizon during the long evening hours of their northern homeland.

The Huang He - China's Cradle of Civilization

With a length of 5,500 kilometres, the Huang He is the second-longest river in China, surpassed only by the Yangtze to the south. Since time immemorial, it has been the principal river in east-central Asia, often referred to as the Cradle of Chinese Civilization as it was along its banks that ancient ancestors of today's Chinese people began to practice agriculture and Chinese civilization first arose.

The Huang He originates at an elevation above 4,600 metres in the Bayan Har Mountains. From its origin, it winds its way through the broad highlands of the Tibetan Plateau and enters a series of steep canyons and cascades until it finally leaves the plateau near the city of Lanzhou in southeastern Gansu province. From Lanzhou, the river enters a vast highland area in north-central China known as the Huangtu Gaoyuan Lo, the world's largest loess plateau. The underlying rock systems of the plateau are covered by a mantle of fine-grained, wind-deposited, yellowish alluvium, known as *loess*, that averages between fifty and sixty metres in depth. Over a time span of millions of years, the river has cut deep valleys with terraced slopes through the loose deposits, carrying away huge quantities of suspended surface material and turning the water a striking yellow colour. Hence the name Huang He, which translates into English as Yellow River.

Upon leaving the Loess Plateau near the city of Zhengzhou, the river slows and broadens onto the North China Plain, over which it flows generally eastward until emptying into the Bo Hai embayment of the Huang He or Yellow Sea. The plain is a large rift basin, formed some twenty-five million years ago by the separation of tectonic plates deep underground, in much the same manner as the Great Rift Valley of east Africa. As the rifting occurred, the land sank, forming a vast basin that became flooded from the sea. Twenty million years ago, the Bohai Bay extended 750 kilometres inland to Zhengzhou. Over the millennia, however, silt deposits from the Yellow River gradually filled the basin, forming the vast, nearly featureless, North China alluvial plain.

In modern times, the North China Plain is one of China's most important agricultural regions, producing corn, sorghum, winter wheat, vegetables, and cotton. The region is referred to as the "Land of the yellow earth," and it is the one of the world's most densely populated regions, being home to mostly Han Chinese, the world's largest ethnic group. The Han Chinese trace a common ancestry to the Huaxia, a name for the initial confederation of two tribes, the Hua and Xia, who first settled along the middle and lower reaches of the Yellow River. The Huaxia were the first to practice agriculture in China, about ten thousand years ago, and it was the Huaxian agricultural confederation that gave birth to Chinese civilization with the founding of the Xia dynasty, the first imperial dynasty in China, four thousand years ago.

From its origins along the middle and lower reaches of the river, Chinese civilization blossomed and subsequently spread across east Asia. The four great inventions of gunpowder, the compass, paper-making, and printing are all contributions to the advancement of human culture that originated in this cradle of civilization. And in Han Chinese culture, yellow is regarded as a colour of ancient origins—an emblem of the Yellow River and the yellow loess land through which it flows. Yellow was the emperor's colour in Imperial China and is held as the symbolic colour of the five legendary emperors of ancient China. It is the colour of the skin of Chinese people and of the legendary Chinese Dragon from which, legend has it, the Chinese people descended.

It was almost two million years ago that *Homo erectus* appears to have reached the Yellow River region. Fragments of bone belonging to *Homo erectus lantianensis*, or Lantian Man, dated to approximately 1.6 million years ago, have been found at sites in Lantian County at the eastern end of the Loess Plateau. Bone fragments and stone tools have been found at Shangchen, also in the Lantian County area, dated to two million years ago. And further south, in Yuanmou County in the southwestern province of Yunnan, bone fragments of Yuanmou Man have been discovered and dated to approximately 1.7 million years ago.

About forty kilometres southwest of Beijing, where the Yanshun Mountains rise from the North China Plain, lies the Zhoukoudian archaeological site where natural limestone caves and freshwater streams provided a habitable environment for early humans for over 750 thousand years. Excavation at the site has uncovered fossils of hominin crania, mandibles, and teeth dating to between 780 and 300 thousand years ago. These specimens show evidence of a progressively larger cranium size from an initial nine hundred cubic centimetres for the older remains to 1,200 cubic centimetres for the most recent. The site has yielded over one hundred thousand stone tool fragments and several hundred animal remains, all of which indicate a more or less continuous occupation by early humans over hundreds of thousands of years. In addition, the remains of what appear to be an archaic *Homo sapiens* subspecies, dating to between two hundred and one hundred thousand years ago, have been found. And there is evidence that *Homo sapiens* from Africa arrived at the site between forty and thirty thousand years ago.

The oldest of the remains, dated to between 770 and 300 thousand years old, have been classified as *Homo erectus pekinensis*, or Peking Man, a contemporary of *Homo erectus heidelbergensis* in Europe and Africa. Evidence indicates, however, that Peking Man evolved in isolation from his western counterparts, likely descending from early *Homo erectus* bands that appeared in eastern China approximately two million years ago.

Peking Man was larger and more robust than humans today and anatomically very similar to Heidelbergensis. In the latter stages of their earthly tenure, they had a cranial capacity ranging from about 1,000 to 1,300 cubic centimetres, a head shape that was flat in profile with a small forehead, a keel along the top for attachment of powerful jaw muscles, very thick skull bones, heavy brow ridges, and a large, chinless jaw. Their teeth were essentially modern, although the canines and molars were relatively large. Their limb bones were somewhat more robust but otherwise indistinguishable from those of a typical modern human.

While the limited archaeological evidence indicates that Peking Man lived until approximately three hundred thousand years ago, the presence of yet-to-be-classified archaic human remains at the Zhoukoudian Cave site, dated to between two hundred and one hundred thousand years ago, suggests that Peking Man continued to evolve in the direction of modern humans. In addition, there is evidence that when early humans from Africa arrived at the site, between thirty and forty thousand years ago, they socially integrated and interbred with the indigenous residents, forming a unique genetic framework from which the modern Asian race is derived.

A discovery of human remains in Tianyuan Cave near the Zhoukoudian site in 2003 does lend credence to an interbreeding theory. The Tianyuan Man skeleton has been radiocarbon dated to forty thousand years ago, and an initial morphological analysis confirmed that it is essentially that of a modern human, with a few archaic traits that could indicate genetic contributions from earlier hominin forms. Then, in 2013, a partial DNA analysis that was performed on the Tianyuan Man skeleton confirmed the morphological analysis, revealing that the individual derived from a population that was ancestral to many present-day Asians, and somewhat surprisingly, Native Americans as well.

Passing the Evolutionary Torch

> *"We used to be hunter-gatherers, now we're shopper-borrowers."*
>
> ...Robin Williams

By the end of their tenure on earth, the erectines had acquired the primary elements upon which modern human culture is founded. They had developed an advanced toolmaking industry, discovered an energy source in the form of controlled fire, and created a cooperative social organization centred around the nuclear family. With their large brains and near-modern anatomy, the erectines had also evolved with at least the beginnings of a capacity for conceptual thought and language. We can never know how sophisticated the erectine language was or how profound their conceptual thinking had become, but the beginnings of these capabilities would most certainly have been present. One can only muse about the feelings they would have harboured and the thoughts they would have pondered as they stared into a campfire, watched a sunset, or gazed at the moon and stars as they coursed their paths across the night sky.

The erectines' cultural inventions marked the beginning of a new phase of human evolution during which the forces of natural selection were tempered by human invention. The erectines affected the direction of their own evolution by creating technological and behavioural solutions to the environmental challenges they encountered, and these achievements reduced the need for evolutionary, genetic responses to those challenges.

Prior to the arrival of the erectines, when animals moved into new environments, the forces of natural selection typically forged evolutionary changes that helped them survive and adapt, altering the population's gene pool in the process. When *Homo erectus* moved into more temperate environments, one would expect such things as increased amounts of insulating body fat and thicker body hair would evolve, but *Homo erectus* adapted by occupying caves or building shelters, using fire for warmth, and using animal hides for clothing. By applying their intelligence and accumulated knowledge, *Homo erectus* began to invent technology and behavioural patterns that enabled them to adapt to new environmental conditions rather than depending on evolutionary changes.

And natural selection only reinforced the trend by continuing to select for a larger brain size and the associated increased cognitive capabilities. As the erectine intelligence increased, their ability to modify their lifestyle and environment increased in concert, and the need for evolutionary adaptation decreased.

This pattern of cultural and technological development trumping natural selection increased dramatically with the subsequent evolution of *Homo sapiens*. Today, most of us live in unnatural, urban environments of our own construct, and the rate of culture change continues to accelerate. We have occupied most environmental zones and land masses on earth, ranging from the high Arctic tundra to the equatorial rain forest, and yet biologically, we are still tropical animals.

We humans today have inherited a body little changed from that of the African erectines, a body beautifully crafted, through millions of years of evolution and adaptation, for life on the tropical savanna. As such, the modern human body is an extraordinary biological machine with capabilities that are unique in the animal kingdom. We can run for hours in forty-plus-degree heat in pursuit of prey, track them down and kill them with skilfully crafted weapons, the product of our powerful intellect. We can binge-feast on meat, rich in energy-producing calories, and our amazing bodies absorb and store those excess calories as fat to be retrieved and converted back into energy when needed for our next long-distance hunt or to see us through those long periods of time when ungulates are scarce and we have to get by on only a handful of calories a day.

Today, however, most of us acquire our calories by visiting a local market where we pick up our ungulate steaks and tubers, taking them home and cooking them on fires started with the turn of a dial. Our steaks are often lathered in calorie-rich sauces, and our tubers are sliced into strips or flakes and deep fried in calorie-rich oils. On such occasions, our calorie intake is far beyond anything that the processes of natural selection could possibly have prepared our bodies for, and we can binge-feast on such foods whenever we please. In anticipation of future periods of time when food will be scarce, our bodies still produce a hunger sensation not long after digesting a big meal, prompting us to eat again while food is available.

Seldom, if ever, however, are we forced to go without calorie-rich food for even a day, let alone weeks. Meanwhile, our bodies do what evolutionary adaptation designed them to do: they absorb all the excess calories we consume and turn them into stores of fat in anticipation of transforming them back into energy when future circumstances require it. That day seldom, if ever, comes, however, but our bodies don't anticipate that and continue to prompt us to keep eating so they can add more and more stores of fat. The result is evident on the streets of almost every city around the world today. It has been estimated that about one-third of the world's human population is now overweight or obese.

It also now appears that human life on earth is entering a third evolutionary phase wherein we humans not only largely control our own evolutionary direction, but are now affecting the evolutionary fate of all life on the planet. Today, human population growth and cultural activity are altering the earth's climate and biospheres at an ever-accelerating rate, previously unseen in the earth's history, and threatening the very survival of all flora and fauna on the planet. Natural restorative processes and evolutionary selective forces simply cannot operate within the accelerated time frames of this new, human induced, environmental change. Only we humans, with our intellectual capacity, scientific knowledge, and technology, are in a position to arrest the catastrophic acceleration of population growth, environmental destruction, and climate change that now threatens our survival and the survival of much of the world's flora and fauna.

Nature, it seems, has handed us the torch when it comes to directing the future course of evolution, making us, like it or not, the custodians of life on Planet Earth. Responsibility for the future well-being of our earth, and the myriad of amazing life forms it supports, now rests squarely on our shoulders.

CHAPTER FOUR

The Sapiens

"We must, however, acknowledge, as it seems to me, that man with all his noble qualities... still bears in his bodily frame the indelible stamp of his lowly origin."

...Charles Darwin

BEGINNING ABOUT THREE hundred thousand years ago, a third generation of erectines began to appear throughout Africa and Eurasia, all of which were anatomically very similar to we humans today. These subspecies were widely dispersed, having descended from *Homo erectus heidelbergensis* in Europe and Africa, from *Homo erectus pekinensis* in eastern Asia, and perhaps from other second-generation erectines yet to be discovered that lived throughout east and southeast Asia between 500 and 250 thousand years ago. In spite of their wide geographical separation, all of these third-generation erectines shared remarkably similar anatomy. They were not yet fully modern, with most being somewhat more robust and heavily built than us, with relatively large heads, elongated skulls, and heavy eyebrow ridges.

Culturally, most third-generation erectines had mastered similarly advanced toolmaking traditions and most, if not all, could make fire on demand. They lived in similar-sized bands with a social organization that was centred on the nuclear family, and they communicated with a rudimentary language with which they organized themselves for hunting, gathering food, and maintaining their communal dwellings. They lived mostly in caves or rock shelters, but when necessary, they constructed shelters of wood, rock, hide, and bone. They stitched clothing from animal hides and cooked their food in firepits. In the cooler climates, they kept their cooking fires burning into the night hours for warmth and light.

While there is an argument for treating these new arrivals as a third-generation *Homo erectus* subspecies, they are now universally regarded throughout the scientific community as members of a separate species, collectively referred to as archaic *Homo sapiens*, or *The Sapiens*.

In recent decades, there have been a large number of discoveries of these sapiens subspecies, and a picture is emerging of a large population of similarly advanced early humans that were widely dispersed throughout Africa and Eurasia, with each appearing to have had the potential to evolve into fully modern humans. A summary of the geographical location of some of the sapiens discoveries is provided in the following Table Three.

LOCATION	TAXONOMICAL DESIGNATION	LOCAL NAME	AGE (KYA)*	CRANIUM VOLUME (CC)**
Europe/Western Asia/Middle East				
Numerous and widely distributed	Homo sapiens neanderthalensis	Neanderthal	400- 25	1600
South and Southeast Asia				
Ngandong, Indonesia	Homo soloensis	Solo Man	550 - 143	1150 to 1300
Madhya Pradesh, India	Homo sapiens narmadensis	Narmada Man	300 - 100	1155 - 1421
Flores Island, Indonesia	Homo floresiensis	Flores Man	190 - 50(18?)	380
Central and East Asia				
Altai Mountains, Siberia	Homo sapiens denisova	Denisovan	400 - 40	
Tibetan Plateau, China	Homo sapiens denisova	Denisovan	160	
Henan Province, China	Homo sapiens denisova?	Archaic Homo	125 - 105	1800
Hexian, China	Homo sapiens transitional	Hexian Man	190 - 250	1025
Dali, China	Homo sapiens transitional	Dali Man	250	1120
Jinniushan, China	Homo sapiens transitional	Jinniushan Woman	260	1330
Zhoukoudian Cave, China	Homo sapiens transitional		200 - 100	
Africa				
Jebel Irhoud, Morocco	Archaic Homo sapiens		350 - 280	1400
Florisbad, South Africa	Archaic Homo sapiens		294 - 224	1400
Omo National Park, Ethiopia	Homo sapiens sapiens	Omo 1	195	
Herto Bouri, Ethiopia	Homo sapiens sapiens/idaltu	Herto Man	160	1450
Laetoli, Tanzania	Homo sapiens sapiens	Ngaloba LH-18	120	1200
*- thousands of years ago				
**-cubic centimetres				

Table Three - Geographic dispersal of archaic *Homo sapiens*

The most well-known of the sapiens are the Neanderthals, Denisovans, and *Homo sapiens sapiens*. Of all the archaic sapiens, *Homo sapiens sapiens* is the only subspecies to have survived into modern times, and all humans living today are their descendants.

The Neanderthals

Homo sapiens neanderthalensis, or the Neanderthals, were widespread throughout most of Europe and west-central Asia beginning about 450 thousand years ago, lasting until roughly forty thousand years ago when they somewhat abruptly disappeared from most of their range.

The subspecies is named after one of the first sites where its fossils were discovered in the nineteenth century—a cave in the Neander valley near Dusseldorf, Germany. Over four hundred sites of Neanderthal remains have since been discovered throughout western and central Europe, the Carpathians, the Balkans, the Ukraine, western Russia, and east to the Altai Mountains and the Indus River.

There is no evidence that the Neanderthals entered Africa, and it is speculated that they may have had difficulty competing with the more aggressive *Homo sapiens sapiens* that occupied east Africa at the time. Whatever the reason, *Homo sapiens neanderthalensis* and *Homo sapiens sapiens* appear to have avoided one another as there is no evidence that they shared the same locale at any time prior to about forty-five thousand years ago when the latter first arrived in Europe from Africa. And that encounter, it seems, did not improve relations between the two because the Neanderthals all but disappeared only a few millennia after the Africans' arrival.

The Neanderthals are believed to have been the direct descendants of *Homo erectus heidelbergensis*, and like their ancestors, they were well adapted to life in cold climates. Neanderthals lived during the last Pleistocene glacial, when icefields often extended well into northern Europe. During the coldest periods, the Neanderthals migrated south into Spain, southern France, Italy, Greece, and around the eastern Mediterranean into the Levant, where evidence indicates that they occupied locales vacated by *Homo sapiens sapiens* bands that had, in turn, retreated back into Africa when the climate cooled. But when the climate turned milder, the Neanderthals again headed as far north as Belgium, England, and Wales.

There is no consensus concerning the population of Neanderthals, partly because they were widely dispersed over a very large territory and partly because their population appears to have fluctuated significantly with the rapid changes in climate that occurred throughout Eurasia during their time frame. Life was harsh for the Neanderthals, and their population was never terribly large. Total population estimates range from as low as ten thousand to as high as seventy thousand individuals during the height of their tenure.

Like their heidelbergensis ancestors, the Neanderthals had physical attributes that would have made their bodies more thermally efficient, enabling them to tolerate the colder climates of Europe and western Asia. They had a stockier, more robust build than we modern humans, with shorter legs and larger torsos. Neanderthals had stronger arms and hands, and their heads were large, with a cranial capacity as much as 1,600 cubic centimetres, much larger than that of today's average human. Their faces were characterized by pronounced brow ridges, reduced chins, wide midfaces, angled cheek bones, and large noses that could have been an adaptation to humidify and warm the cold, dry air of northern climates.

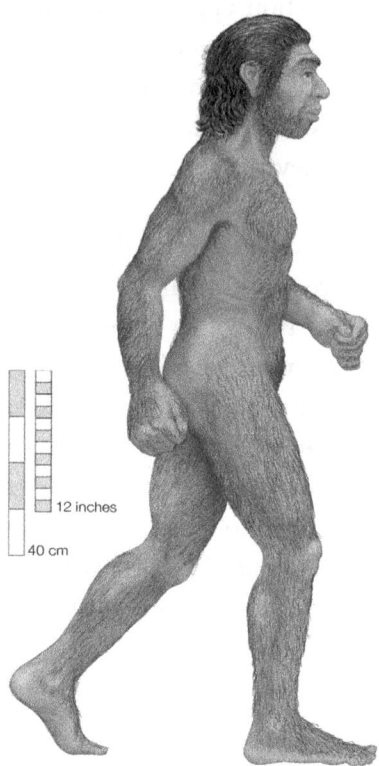

An artist's rendering of *Homo sapiens neanderthalensis*.

© *2005 by Encyclopædia Britannica, Inc.*

There is evidence that Neanderthals may have been capable of sophisticated speech. In human anatomy, the hyoid bone, situated centrally in the upper part of the neck just above the larynx, provides the anatomical foundation for speech. In 1989, a sixty-thousand-year-old Neanderthal hyoid bone was discovered in the Kebara Cave in Israel, which, in terms of mechanical behaviour, was basically indistinguishable from that of a modern human. Hence, it would appear that the Neanderthals were at least anatomically as capable of speech as we are.

The Neanderthals had an omnivorous diet that varied significantly, depending on the environment in which they lived. An analysis of the dental tartar on the teeth of Neanderthal remains, recovered from El Sidrón Cave in Spain, revealed a large amount of plant material such as nuts, moss, and mushrooms. A similar analysis from teeth recovered from Spy Cave in Belgium indicated a meat-based diet, including woolly rhinoceros and wild sheep.

For the most part, however, Neanderthals were meat eaters. Evidence shows they were proficient big-game hunters and preyed upon the wide variety of large mammals that were found throughout Eurasia during their time.

By one hundred thousand years ago, possibly somewhat earlier, Neanderthal toolmaking had significantly advanced over the Acheulean tradition of their erectine ancestors. Their toolmaking tradition, known as *Mousterian*, used a more complex, three-step stone-knapping process called the Levallois technique, which produced flakes, rather than cores, as the primary end product. This flaking technique required the careful selection of a flint or quartz cobble of the approximate size and general shape of the desired end product, and typically, a single flake tool was made from each cobble.

To craft a Levallois tool, a flat striking surface or platform is first formed at one end of the cobble stone by striking it in such a way as to leave a flat end. Next, the edges of one side are trimmed by flaking off pieces around the entire outline of the intended flake, creating a domed shape on the side of the core. The result is known as a tortoise core as the various scars and rounded form are reminiscent of a tortoise's shell. Finally, the striking platform is struck at a precise angle, and a flake separates from the core with all of its edges sharpened by the earlier trimming work.

The Levallois technique produced thinner flakes with sharper points and razor-sharp edges, and it allowed for finer control over the size and shape of the resulting points and tools. Neanderthals used the Levallois technique to produce tools in a wide variety of shapes and sizes, and many were fabricated with a specific application in mind.

A Levallois flint point from Syria.

    The Neanderthals were proficient big-game hunters, their culture was centred around socialized hunting practices, and animal products were incorporated into most aspects of their daily activity. In addition to a primarily meat diet, they made clothing from animal hides, and when caves were not available, they made shelters with mammoth bone frames covered with the hides of the enormous beasts.

    The Neanderthals fabricated large, hafted spears that provided much better penetration of animal hides, enabling them to slay larger mammals such as cave bear, auroch, rhinoceros, horse, ass, bison, elk, and even woolly mammoth. And with their Mousterian technology, they could produce sharper blades and hand axes that were better suited for butchering these large animals. Other tools were fabricated specifically for scraping hides and preparing them for use as clothing and shelter covers. Hides were stitched together with sinew using awls of wood or bone.

The Neanderthals sometimes constructed open-air camps with shelters made of animal bones and hides as evidenced by discoveries at a Neanderthal archaeological site in the Dniester River valley in the Chernovtsy Province of Ukraine. The site has yielded some forty thousand lithic artifacts manufactured in the Mousterian tradition along with about three thousand mammal bones, mostly from mammoths. The site contained activity areas for butchering and tool production, twenty-five hearths, and a circular accumulation of mammoth bones, deemed to have been the foundation and framing over which stitched hides were draped to create a dwelling. Based on the size of the site and the number of artifacts found, it appears that it had been used for some time by a relatively large number of individuals. The site was the exception, however, as most Neanderthal archaeological sites found throughout Europe and western Asia have been in caves and rock shelters and indicate that group sizes were typically no more than about thirty individuals.

The Neanderthal life was hard, not only because of the harsh climate and challenging environment in which they lived, but also because their hunting practices were physically demanding and dangerous. Given their short legs and heavy build, and the less-open terrain in which they lived, Neanderthals were not persistence hunters like their sapien cousins in Africa. They typically ambushed their prey in forested terrain or trapped them in ravines, canyons, or marshes—places from which large animals would have difficulty escaping. There is no evidence that the Neanderthals had projectile weapons such as the bow and arrow or atlatl, but rather, they hunted in groups and relied on getting close enough to an animal to kill it with thrusting spears, stone clubs, and knives.

When an animal was chased into a ravine or some form of natural trap, it would most certainly have turned and charged its attackers in a desperate attempt to escape. Typically, the group of Neanderthal men, women, and children would surround the desperate animal—an auroch, cave bear, or mammoth—and thrust multiple spears and knives into its flesh. At the same time, they would try to avoid the thrashing horns, tusks, jaws, hooves, or claws of the animal until the animal finally succumbed to its wounds or the Neanderthal band had suffered too many injuries and was forced to withdraw.

It is no surprise that the fossil evidence indicates that Neanderthals regularly sustained serious injuries while killing large, wild animals, and it is likely that death or permanent crippling from such injuries was common. Killing large animals would, no doubt, have often been a long, drawn-out battle, accompanied by injury and even casualties.

In the sites where numerous Neanderthal remains have been found, many show evidence of healed bone fractures on limbs, torso, or skull, and many individuals had been permanently deformed by injuries sustained at a young age. There is also evidence that Neanderthals suffered from a wide range of ailments, including pneumonia and malnourishment. Few Neanderthals would have died from old age. Most died from predation, injury, disease, or starvation, and few would have lived beyond the age of thirty years.

Neanderthals appear to have been a compassionate people, as evidence shows that they took care of each other while injuries healed or when elders became too old and infirm to take care of themselves. The skeleton of one Neanderthal man found in a cave near Shanidar, Iraq, showed evidence that the man had suffered crushing injuries early in life and sustained multiple broken bones. By the time he died, his injuries had led to degenerative joint disease, the withering of one of his arms, and blindness in one eye. Such injuries would have taken some time to heal and would have precluded the man from participating in a hunt, making tools, or even helping out around camp for the rest of his life. Yet it appears that he lived for many years after his injuries. In the same cave, researchers found the remains of an individual that had lost all teeth and suffered from severe arthritis, and yet it had lived to the ripe old age of forty-five years. In such cases, the individuals could have survived only if they had been cared for by family or other group members for many years.

It also appears that Neanderthals practiced ceremonial burial as several Neanderthal cave sites provide strong evidence of intentional burial by ninety thousand years ago. Buried remains, typically flexed into a fetal position, have been found in shallow graves dug into the soft midden soil at the mouths of caves and rock shelters. Frequently, the buried bones were stained with hematite, a rust-red iron ore, likely either sprinkled on remains as a powder, or painted on the bodies after the powdered pigment was mixed with water or liquid animal fat. In nearly half of the thirty-three known Neanderthal burials, stone tools and animal bones were also found in the graves, and in at least one case, the body of a man was placed on a bed of pine boughs, and eight species of flowers, including hyacinths, daisies, hollyhocks, and bachelor's buttons, were sprinkled over him.

One can only speculate on the significance of such burial rituals in terms of religious belief or spirituality. Some regard deliberate burial as a sign of belief in an afterlife, but no other religious artifacts have been found to support such a hypothesis. Rather, the burial practices were more likely simply a sign of emotional attachment to the deceased and demonstrate a desire to treat the departed's remains with dignity and respect. Such burial practices are, however, evidence of emotional bonding, the beginnings of a sense of humanity, and an awakening spirituality.

By one hundred thousand years ago, the Neanderthals were at the apex of their tenure on earth. Life was by no means easy, but they had adapted to the cold climate in which they lived, and they thrived in spite of the hardships they faced. But it was about this time that the last of the Pleistocene glacial maximums, commonly referred to today as the Great Ice Age, began to intensify. Average global temperatures were steadily falling, reaching their lowest point between twenty-five and thirty thousand years ago.

As the cold grip of the advancing ice age spread across the globe, massive icefields formed across northern Eurasia and all of Canada, reaching as far south as the northern United States. However, climate conditions left much of Europe and central Asia free of ice, even though the temperatures were extremely cold. With so much of the earth's water locked up in glaciation, rainfall diminished, and sea levels fell. The climate globally became cold and dry; throughout Europe, forests and woodlands shrank into a few river valleys and isolated pockets in the mountain ranges in the south. In their place, an enormous grassland steppe formed that covered most of northern and central Eurasia, stretching from the Iberian Peninsula across Siberia to Alaska.

In spite of the harshness of the climate, the grasslands of the steppe were productive, and the steppes teamed with exceptionally large mammals, known as *Pleistocene Megafauna*. Some of the more notable of these giants were multiple species of mammoth, straight-tusked elephant, auroch, bison, deer, and rhinoceros. The top predators of the day were the sabretooth cat, the cave lion, the cave bear, the leopard, and the Neanderthals.

As the forests and woodlands retreated southward, Neanderthal bands either retreated with them or adapted to life on the steppe grasslands. Lacking projectile weapons and being less capable runners than their sapien cousins in Africa, they were less successful at obtaining meat on the open steppe lands and preferred to live in river valleys and forested areas along the edges of the steppe. In these locations, they could hunt both the steppe megafauna and woodland game. But despite their ability to hunt successfully in such locations, by about fifty thousand years ago, the Neanderthal population throughout Eurasia was starting to decline. And while climate change was a significant factor, the arrival of bands of African *Homo sapiens sapiens* no doubt contributed to the demise of the Neanderthals as well.

By about fifty thousand years ago, the African Homo sapiens sapiens were socially and culturally much more advanced than the Neanderthals with essentially modern anatomical and behavioural characteristics. These later Homo sapiens sapiens are referred to as *Modern Humans* by many paleoanthropologists to distinguish them from earlier, anatomically modern but culturally less advanced archaic Homo sapiens sapiens that were living throughout east Africa prior to this time.

About forty-five thousand years ago, bands of Modern Humans began to appear in southern and central Europe, having migrated out of Africa via the Levant. By this time, the global population of Modern Humans had grown to several hundred thousand, living in bands of several dozen individuals. Their culture had likewise blossomed with advancements in social organization, artistic expression, and technology that included throwing spears and possibly the bow and arrow. These new immigrants were well equipped for life on open grasslands as, apart from the extreme temperature difference, the steppes were not that different than the savanna of east Africa. Bands of Modern Humans appear to have thrived on the steppes of Europe and were formidable hunters of the megafauna that lived there.

As these larger bands of Modern Humans fanned out across Europe and west-central Asia, they encountered Neanderthal bands that were much smaller in numbers and culturally less advanced. The Neanderthals had no answer to the superior technology and advanced hunting skills of these new foreigners who quickly outcompeted them for the best habitats and most accessible animal resources. The Neanderthals became marginalized and were forced to retreat to less hospitable environments and harsher living conditions. As a result, their numbers plummeted.

This does not mean that there was no social interaction between the human newcomers and the Neanderthals. Quite the contrary. Over the last few decades, many hundreds of complete or partial skeletal Neanderthal remains have been discovered, and scientists have been able to extract enough intact DNA from bone fragments to map the Neanderthal genome, which was published in 2010. A comparison of the Neanderthal with today's human genome reveals that on average, about 3 to 4 percent of the DNA in Europeans today is of Neanderthal origin, indicating that the two species interbred.

Nevertheless, within five thousand years of the arrival of Modern Humans, the Neanderthals had been pushed to the brink of extinction throughout most of Eurasia. Only a small number survived in a few southern refuges where the environment was more hospitable and where there were few, if any, Modern Humans present. One such southern refuge was the area today known as the Rock of Gibraltar.

The Rock of Gibraltar is a massive limestone peninsula that juts out into the Mediterranean Sea, just east of the western entrance from the Atlantic Ocean. The peninsula is riddled with more than a hundred caves, and on the southeastern side, there are a series of caves that extend deep into the limestone cliffs with the Mediterranean Sea lapping just below their entrances. Evidence unearthed from these caves shows that they were occupied by Neanderthals starting about one hundred thousand years ago, and they reveal a wealth of information about their lifestyle and the prehistoric landscape of the area.

During the last Pleistocene Ice Age, the geography of the Gibraltar region looked very different than it does today. With the large volume of earth's water locked up in the polar ice caps and continental glaciers, the sea level was as much as eighty-five metres lower, exposing a coastal plain that extended some four kilometres eastward from the base of the caves. On the western side of the peninsula, today's Bay of Gibraltar was a marshy estuary, and to the north was the western edge of the vast steppe grasslands.

The coastal plain and estuary were home to a wide variety of fauna, including deer, ibex, wild cattle, rabbits, and boar that were preyed upon by hyena, lynx, wolves, bears, and various species of large cats. Animal remains found in the caves show that the Neanderthals were active hunters and had a preference for ibex and deer as both are well represented in the remains found there. In addition to large animals, they also appear to have enjoyed large quantities of small mammals and birds, particularly rabbits and migratory birds, both of which must have been abundant in the coastal marshlands.

The Neanderthals also appear to have harvested food from the sea as many mussel shells, along with the bones of sea mammals, have been found in the caves. Circumstantial evidence suggests they may even have built watercraft and paddled along the northern coast of the Mediterranean Sea as early as one hundred thousand years ago, which would make them among the earliest known seafarers in the archaeological record.

During the time of the Neanderthals, the Gibraltar region enjoyed a much more stable and temperate climate than almost anywhere else in Europe. Whereas northern Europe underwent massive climatic swings between temperate and extreme glacial conditions, which made large areas of the continent uninhabitable for extended periods, Gibraltar's prehistoric climate appears to have been only mildly affected by such changes. As a result, it became a kind of Africa in Europe where animals, plants, and Neanderthals were able to find shelter from the worst effects of the Great Ice Age.

Between 25,500 and 22,500 years ago however, the earth was plunged into a deep freeze, the apex of the Great Ice Age, and the climate of the Gibraltar region became extremely cold, arid, and unstable. The abrupt climate change would have greatly disrupted the Gibraltar Neanderthals' food supply, and it appears to have stressed their population beyond recovery, leading to their final extinction about twenty-five thousand years ago.

There is no archaeological evidence indicating that the Neanderthals experienced a cultural explosion like their sapien relatives in Africa. While the Modern Humans of Africa were exhibiting artistic expression, advanced weapon technology, and more advanced social organization by fifty thousand years ago, the Neanderthal culture remained essentially unchanged during their entire tenure on earth. Their artistic expression showed no sign of figurative art. While they may have carved personal adornment or statues of wood or other perishable materials that would not survive into modern times, the only surviving artistic artifacts are primitive rock carvings on cave walls and possibly crudely carved bone, although the latter is questioned.

Given that the Neanderthals' brain size was at least equivalent to Modern Humans, their lack of more advanced cultural development remains something of a mystery. Some scientists believe it indicates an inferior intelligence, while others argue that their lives were simply too difficult to allow enough free time for cultural activity. Another reason may have been their small population size. Cultural development goes hand in hand with population growth, and since the Neanderthal population remained small throughout their entire existence on earth, the size of individual bands being typically between twenty and thirty individuals, their relatively small group size was no doubt a limiting factor in their cultural advancement.

Solo Man

Homo sapiens soloensis, or Solo Man, was a sapien subspecies that lived on the Indonesian island of Java from about 550 to 143 thousand years ago. Solo Man is known from eleven fossil skulls and two leg-bone fragments that were discovered during excavations in 1931–32 along terraces of the Solo River, from which their name is derived. Solo Man had a cranial capacity of 1,000 to 1,250 cubic centimetres, and their toolmaking capabilities were relatively advanced, but their skull was typical of an erectus-sapiens transitional subspecies, flattened in profile, with thick bones and heavy brow ridges that formed a torus.

Some debate exists within the archaeological community about the classification of Solo Man and their position in the tree of human evolution. Given that their culture appears relatively advanced and more or less aligned with Neanderthals and early Modern Humans, most classify them as the subspecies *Homo sapiens soloensis*. However, given their age and general morphology, which is closer to *Homo erectus*, others refer to Solo Man as the subspecies *Homo erectus soloensis*.

Taxonomy aside, Solo Man appears to have become extinct long before the arrival of Modern Humans into southeast Asia, so it is most likely that Solo Man was simply a second- or third-generation *Homo erectus* that lived for a period of time in isolation from the rest of their species until their disappearance, some 140 thousand years ago. The name Solo Man, it seems, is a double entendre.

Flores Man

Homo floresiensis, or Flores Man, is an intriguing species of hominin that appears to have lived exclusively on the Indonesian island of Flores between about 190 and 50 thousand years ago, with some tentative evidence indicating that a small population may have survived to as recently as eighteen thousand years ago. Flores Man averaged just over one hundred centimetres tall and had a brain size that averaged only about 380 cubic centimetres, about the size of a modern great ape's brain, and remarkably small for a hominin of that time period.

Notwithstanding this small brain size, *Homo floresiensis* appears to have been quite intelligent and culturally advanced. In the caves where they lived, evidence indicates the use of fire for cooking and butchered animal bones. Flores Man's stone tools were comparable to those of the Neanderthals, and given that they successfully hunted and butchered large elephant-like prey called *Stegodon*, Flores Man must have lived in social groups and practiced fairly advanced cooperative hunting techniques.

Because deep waters surround the island, Flores Man remained isolated during the Pleistocene interglacials, despite the low sea levels at the time. This circumstance has led some scientists to conclude that Flores Man or their ancestors could have reached the isolated island only by water transport, perhaps arriving in bamboo rafts around one hundred thousand years ago. Another possibility is that they had simply been washed out to sea in a storm and drifted across the strait on mats of floating vegetation.

Evidence indicates that Modern Humans arrived on the Island of Flores about fifty thousand years ago, around the time that Flores Man and other large animals such as the *Stegodon* essentially disappeared. As with the Neanderthals, Modern Humans were likely a factor in the demise of *Homo floresiensis*.

The origin of Flores Man is unknown, but most paleoanthropologists believe the species evolved from *Homo erectus* and speculate that their diminutive size was the result of a genetic disorder or environmental factors. However, new research by Dr. Debbie Argue from the Australian National University in Canberra suggests that *Homo floresiensis* may, in fact, be a descendant of *Homo habilis* or a closely related Australopithecine. The hypothesis has merit as *Homo floresiensis* is much closer anatomically to the Australopithecines than *Homo erectus* and had a similar brain size. It raises the intriguing prospect that, like *Homo erectus*, *Homo habilis* or a close Australopithecine relative, may have left Africa some two million years ago, or even earlier, and somehow wandered as far as southeast Asia where it thrived and continued to evolve until relatively recent times.

The Asian Sapiens

Over the last few decades, excavations at sites throughout central and eastern Asia have discovered a large number of archaic *sapiens* remains, indicating that there were likely as many sapiens living in the east as in the west during the period from about three hundred thousand to thirty thousand years ago. Although the eastern and western sapiens had evolved in isolation over the course of almost two million years, they were anatomically very similar, and as genetic evidence now reveals, they socialized as members of the same species whenever they encountered one another.

The anatomical similarities between the eastern sapiens and the Neanderthals has prompted some paleoanthropologists to propose that the eastern sapiens should also be classified as Neanderthals. However, the Eastern sapiens are, for the most part, more heavily built than even the

Neanderthals and had somewhat larger but less elongated heads. These differences wouldn't warrant their classification as separate species, but based on such differences, along with recent genetic analysis, most paleoanthropologists now refer to the eastern sapiens as separate sapien subspecies.

The Denisovans

In spite of their location at the geographical centre of the Eurasian land mass, the Altai Mountains of central Asia remain a remote wilderness region with some of the most beautiful scenery in the world. Their snow-capped peaks, high mountain lakes, waterfalls, and turquoise-coloured rivers have awed human travellers for millennia.

The Altai Mountains rise from the Eurasian Steppe, with the Mongolian desert to the east and the grasslands of Kazakhstan and Russia to the west, and the region is often regarded as the place where East meets West, a natural boundary between eastern and western people, culture, and civilization.

Since the third millennium BCE, merchants from Tibet and China have travelled through the region along the ancient Steppe Route to trade with Eastern Europeans, their caravans laden with Persian rugs, silk, furs, precious stones, and jewels. The Steppe Route extended some ten thousand kilometres, from the mouth of the Danube River on the Black Sea to the East China Sea, connecting eastern Europe with northeastern China. Much of the route followed what was once the paths of migrating animals and, over time, became known as the Great Silk Road.

Just north of the ancient trade route, about thirty metres above the bank of the Anuy River in the Altai Mountains, there is a large limestone cavern that has been home to hominins for hundreds of thousands of years. The site is known as Denisova Cave, named after the hermit Dyonisiy, or "Denis," that lived in the cave during the eighteenth century. Excavations in the cave have revealed that it was also home to a sapiens subspecies that lived there from about 280 to 30 thousand years ago. The subspecies has been classified as *Homo sapiens denisova*, or the Denisovans, a name derived from the name of the cave and the hermit that once lived there.

Little is known of the precise anatomical features of the Denisovans since only a small quantity of their bone fragments have been found. However, a Denisovan finger bone found at the site is unusually large and robust, and in 2016, two fragments of a palm-sized section of a braincase were discovered in the cave, which were also extraordinarily thick. These fragments suggest that the Denisovans were much bigger and stronger than Modern Humans.

In 2010, geneticists extracted sufficient DNA material from the finger bone and a tooth to complete the mapping of the Denisovan genome. It was found to be genetically distinct from both Neanderthals and Modern Humans. Surprisingly, the DNA analysis revealed that the finger bone belonged to a female, reinforcing the notion that the Denisovans were very robust and heavily built people, perhaps even more so than the Neanderthals.

During the latter stages of the Denisovan habitation, a period from roughly one hundred to forty thousand years ago, bands of Neanderthals appeared in the Altai Krai region and time-shared the cave. The nature of this time-sharing agreement is unknown, but an analysis of the Denisovan genome has revealed it to contain roughly 17 percent Neanderthal DNA, so the time-sharing arrangements must have been fairly amicable.

In 2018, a remarkable discovery was made by a team of palaeogeneticists from the Max Planck Institute for Evolutionary Anthropology in Leipzig, Germany, when they conducted a genome analysis on six bone fragments recovered from Denisova Cave. The fragments were all from the same teenaged female sapien, who died about ninety thousand years ago, and all contained roughly equal amounts of DNA from a Neanderthal and a Denisovan. The analysis revealed that the young woman was the first-generation offspring of a Neanderthal mother and a Denisovan father. It was the first time that scientists have identified an ancient sapien individual whose parents belonged to separate subspecies.

While the Denisovan DNA analysis confirmed interbreeding, it also created a puzzle. Denisova Cave is located just seven hundred metres above sea level, but the DNA analysis revealed a gene variant that is known to protect against hypoxia (oxygen deficiency) at high altitudes. A possible explanation for the presence of the gene came to light, however, with the discovery of a 160-thousand-year-old mandible fossil in Baishiya Karst Cave high on the Tibetan plateau in China's Gansu province. The fossil was found by a monk who visited the cave in 1980, but it was not until 2018, when scientists could analyze proteins found in one of the molars, that the true significance of the find was revealed. A comparison of the genetic material from the Tibetan mandible fossil with that of the remains from Denisova Cave revealed that the Baishiya Karst Cave mandible belonged to a Denisovan. With a powerful jaw and unusually large teeth, which was consistent with the remains from Denisova Cave and those found elsewhere throughout China, this evidence lends credence to the theory that, like the Neanderthals in Europe, the Denisovans were widespread throughout China and southeast Asia.

The dating of the Baishiya Karst Cave mandible also suggests that the Denisovans had adapted to extremely cold climates as it is known that at the time, temperatures on the plateau were even harsher than they are today, when temperatures can plunge to −30°C. It is possible that the Denisovans lived on the Tibetan Plateau prior to entering the Altai Krai region, which would explain the presence of the high-altitude gene in the Denisova Cave and the residents' tolerance of cold temperatures. Denisova cave, with an average temperature of zero degrees would have seemed quite comfortable to someone adapted to life on the Tibetan Plateau.

The Baishiya Karst Cave discovery showed that as early as 160 thousand years ago, Denisovans had genetically adapted to high-altitude, low-oxygen environments. What is perhaps even more remarkable, the genetic analysis revealed that when Modern Humans arrived in the region about forty thousand years ago, the two subspecies interbred, and the high-altitude, cold-climate gene adaptations were passed down through subsequent generations and are carried by people living in the region today. It seems that the modern-day Sherpas have ancient Denisovan ancestors to thank for their remarkable ability to function in high-altitude, low-oxygen environments without displaying symptoms of hypoxia.

Evidence of Denisovan interbreeding was also found in the genome of modern people living in Papua New Guinea, Melanesia, where between 4 and 6 percent of their genome was found to be derived from a Denisovan population. The Denisovan components were related to brown skin, hair, and eyes, which are all consistent with features found amongst Melanesians today.

Still other genetic studies have shown that the Denisovans may have reached as far south as Australia, as comparative genetic tests indicate possible DNA sharing between the Denisovan and the Aboriginal Australian genome. If true, the results are somewhat surprising as such a journey would have involved significant travel distances by sea, even when sea levels were at their lowest.

From these findings, it appears that, at some point prior to fifty thousand years ago, Denisovan bands migrated into southeast Asia and interbred with Modern Humans that entered the area about forty-five thousand years ago. Given that the Denisovans once lived on the Tibetan Plateau and wandered as far as southeast Asia from their confirmed location in the Altai Mountains, it is likely that they wandered throughout eastern Asia as well. Some paleoanthropologists are beginning to suspect that other hominin discoveries throughout east Asia may also be Denisovans, although to date, no DNA evidence from these other discoveries confirms or refutes such a hypothesis.

The Lingjing People

Approximately four thousand kilometres east of Denisova Cave, near the village of Lingjing, China, there is an open-pit archaeological site that has yielded a rich trove of hominin fossils and artifacts dating from 125 thousand years ago, to the beginning of the Bronze Age. The site is well stratified, and an analysis of the sandy sediment of the deepest layer has revealed that some one hundred thousand years ago, the site was located at the shore of a lake, surrounded by forested hills and freshwater springs and streams. There was abundant flora and fauna in the region, and the human residents were skilled hunters who lived at their lakeside camp for a long period of time.

Some forty-five fossilized fragments that fit together to form two partial crania of hominin skulls have been found at Lingjing. The crania are between 125 and 105 thousand years old, and one of them has an astounding volume of 1,800 cubic centimetres, one of the largest hominin crania ever discovered. Excavations have also brought to light almost six thousand stone artifacts, several hundred identifiable animal fossil specimens, and over one hundred bone scrapers, points, and blades. The stone artifacts included hammer stones, axes, chisels, blades, and points knapped from white vein quartz pebbles and various quartzite blocks, which most likely came from the boulder bed of an ancient river that ran about seven kilometres northwest of the site.

The quality of the artifacts associated with the Lingjing site indicate a relatively advanced weapon and toolmaking tradition for the time, fabricated specifically for the killing of large animals. Faunal remains found at the site include extinct large herbivores like the straight-tusked elephant, woolly rhinoceros, auroch, and the giant deer *Megaloceros ordosianus*. The people that killed them must have been skilled hunters with lethal weaponry and advanced cooperative hunting practices, hardly surprising given the 1,800 cubic centimetres of grey matter volume of at least one of the residents.

Dali Man and the Jinniushan Woman

In 1984, the 260-thousand-year-old fossilized remains of an archaic sapien were excavated at the Jinniushan archaeological site, a collapsed limestone cave in the Liaoning province in northeast China. In all, the remains consisted of a skull, left ulna, six vertebrae, ribs, and numerous bones of the hands and feet, all of which were found to belong to a single, female individual. The head was elongated with a relatively thin cranial vault bone with heavy brow ridges and a cranial capacity of approximately 1,400 cubic centimetres.

The Jinniushan woman had an exceptionally large, robust body. She had a wide trunk and short limbs, an adaptation to life in colder climates and consistent with other sapiens discovered throughout China. Jinniushan Woman's body mass has been estimated to have been approximately seventy-nine kilograms, making her the largest female hominin specimen discovered to date.

A few hundred kilometres to the southeast of the Jinniushan site, in Dali County, the fossilized remains of another archaic sapiens, known as

Dali Man, was discovered and also dated to 260 thousand years ago. The specimen consisted of a nearly complete fossilized skull that is very similar in appearance to that of Jinniushan Woman, being low and long with a rounded posterior end and heavy brow ridges. The cranial volume of the skull was only 1,120 cubic centimetres, however, indicating that Dali Man was perhaps not the sharpest flake in the debitage pile. He probably would have relied on Jinniushan Woman to do most of the heavy thinking, and given her rather formidable size, it was probably best that Dali Man defer decision-making to Jinniushan Woman in any event!

As more discoveries of hominin remains are made throughout Asia, all sharing similar anatomical characteristics, the case for a single East Asian sapiens subspecies continues to gain support. Like the Neanderthals to the west, it is beginning to appear that a single sapiens subspecies, which many refer to as the Denisovans, occupied a vast territory that ranged from the Altai Mountains in the west, across China to the Pacific Ocean, and south through southeast Asia as far as Australia.

The Neanderthals and the Denisovans evolved in isolation from one another but were remarkably similar anatomically, and when they did meet, the meetings appear to have been amicable. They most likely socialized as one human species, coexisted as one people, and paired and raised mixed families. However, while they appear to have been evolutionarily poised to do so, neither the Neanderthals nor the Denisovans evolved further into fully Modern Humans. To date, all evidence points to the likelihood that Modern Humans first appeared in Africa and evolved from a member, or members, of the African Transition Group.

The African Sapiens

On the African continent, there have been several discoveries of archaic *Homo sapiens* remains dating to between roughly four hundred and one hundred thousand years ago, found at widely separated sites from Morocco in the far northwest, throughout the east African Great Rift Valley system, and as far south as Free State Province, South Africa. For the most part, these African subspecies were anatomically similar to the Neanderthals and Denisovans and are sometimes collectively referred to as the African

Transitional Group, or ATG. The two oldest members of the ATG have been found at the Jebel Irhoud site, located just west of Marrakesh, Morocco, and at the Florisbad site in Free State Province, South Africa.

The Jebel Irhoud site contains the remnants of a karst cave that was filled with deposits laid down in layers during the Pleistocene, preserving many flint blades, points, knives, scrapers, drills, and other tools made in the Mousterian tradition. The site also contained the remains of several sapien individuals dating to between 350 and 280 thousand years ago. The human remains found at the site include part of a skull, a jawbone, teeth, and limb bones that came from three adults, a juvenile, and a child aged about seven years old. The bones indicate that the Jebel Irhoudian people had large, elongated heads with heavy brow ridges, relatively massive jaws, and elongated faces similar to the Neanderthals—so similar, in fact, that some paleoanthropologists have speculated that the sapiens at Jebel Irhoud may have been Neanderthals. Given that the Jebel Irhoud site is not far from the Straits of Gibraltar and that the European continent is visible only a few kilometres away from the northern tip of Morocco, the possibility that early sapiens crossed the strait from time to time, in either direction, can't be readily dismissed.

The Florisbad remains, sometimes referred to as *Homo sapiens helmei*, consist of skull fragments, facial and jaw bones, and a single enamelled molar, which was dated to approximately 259 thousand years old. Analysis of the skull has revealed that the Florisbad sapien was similar in size and anatomy to those at Jebel Irhoud and had a similar brain volume of about 1,400 cubic centimetres, somewhat larger than that of a Modern Human but consistent with the size of most other contemporary sapiens found throughout Africa and Eurasia. The stone tools found with the human remains were also similar to those found at Jebel Irhoud and elsewhere throughout Eurasia for the same time period.

As interesting as the remains from Florisbad and Jebel Irhoud are, however, the most remarkable set of sapien remains was discovered in the Ethiopian section of the Great Rift Valley known as the Afar Depression.

The Arrival of Homo sapiens sapiens

"The greatness of humanity is not in being human but in being humane."

...*Mahatma Gandi*

The Afar Depression is a triangular-shaped lowland that lies at the northern end of the Great Rift Valley in north-central Ethiopia. Geologically, the Afar Depression is unique in that it is located at the intersection of three tectonic plates, which have been spreading apart for millions of years, creating the Afar lowlands as well as the Red Sea and the Gulf of Aden. Geologists refer to the region as the *Afar Triple Junction*.

The region is bordered to the west by the Ethiopian Plateau escarpment, to the east by the Ahmar Mountains, and to the north by the highlands of Djibouti, Eritrea, and Somalia, which separate the Afar lowlands from the shores of the Red Sea. In modern times, the climate in the region is extremely hot and dry, and the Danakil Desert, located in the depression at the borders of Eritrea and Djibouti, is one of the hottest, driest places on earth with daytime temperatures routinely surpassing 50°C, and the annual rainfall is typically less than three centimetres.

The Awash River snakes along the valley floor, providing life-giving fresh water to the flora, fauna, and human inhabitants of the Afar Depression. The river's course is entirely contained within the boundaries of Ethiopia, and it never reaches the sea. The river rises south of Mount Warqe in the Ethiopian Highlands and flows some twelve hundred kilometres down the escarpment, along the valley floor of the Afar Depression, and through a chain of salt lakes to Lake Abhe Bad, a landlocked lake near the border with Djibouti.

The Ethiopian escarpments and Afar Depression have proven to be rich sources of the very earliest hominin remains and are one of the more important paleontological sites in Africa. The region was home to both Ardipithecines, and Australopithecines, including Ardi and Lucy. The erectine Turkana Boy was discovered along the shores of Lake Turkana, located at the southern end of the region, straddling the border between Ethiopia and Kenya.

In 1967, along the Omo River, which flows from the Ethiopian Highlands into Lake Turkana through what is known as the Kibish Formation, Dr. Richard Leakey and a scientific team from the Kenya National Museum found the remains of another, rather remarkable member of the African Transition Group. The human fossils recovered at the site belong to a single individual, referred to as Omo 1, and include a skull, several pieces from the upper limbs, a shoulder, the right hand, legs, pelvis, ribs, and vertebrae, allowing paleontologists to reconstruct a fairly accurate model of Omo 1's anatomy and, in particular, its head shape. Omo 1 is estimated to have been about twenty years of age at the time of death, and although it cannot be known for certain, evidence indicates the remains are those of a female.

Omo 1 had anatomical features very similar to other sapiens subspecies of the time, but unlike the Neanderthals, Denisovans, or other member of the ATG, her head shape was essentially modern. In particular, Omo 1's head was more globular than her sapien contemporaries, less protruding along the lower posterior portion, a higher central and posterior dome, a more prominent jaw, a more vertical forehead, and less protuberant bone ridges above the eyes. In life, Omo 1 would have stood about 175 centimetres tall and weighed about seventy-three kilograms, and she was anatomically modern in every respect. If she was seen walking the streets of Addis Ababa today, she would appear to be a typical, albeit large, African woman.

The Omo 1 remains were initially dated to 130 thousand years ago, but through the application of more sophisticated dating techniques in 2005, the remains have been more accurately dated to between 194 and 196 thousand years old, making them the oldest remains of anatomically modern humans discovered to date. It is on the basis of this discovery that most scientists support the hypothesis that our immediate ancestors, *Homo sapiens sapiens*, first appeared on earth about two hundred thousand years ago in the Afar region of the Great Rift Valley in East Equatorial Africa.

The stone artifacts found in association with Omo 1 were similar to those found with other sapiens. They were manufactured in the Mousterian tradition and included Levallois cores, hand axes, flakes, scrapers, knives, and points of chalcedony and chert. Several non-human bone artifacts, including a variety of vertebrate fossils, dominated by birds and bovids, were also found at the site.

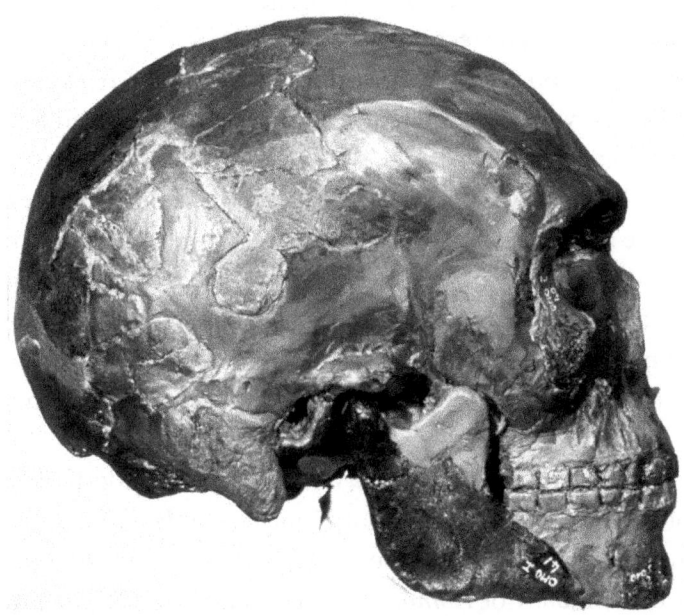

The reconstructed skull of Omo 1 in the Natural History Museum, London.

© *Natural History Museum, London / Science Photo Library.*

In 1997, a team from the University of California, Berkeley, unearthed the largely complete skulls of two more anatomically modern adults and a child at Herto, in the Middle Awash region of the Afar Depression. Dated to approximately 160 thousand years ago, the skulls are very similar to Omo 1 and have been labelled *Homo sapiens idaltu*. And further south, along the Great Rift Valley, in the Laetoli region of Tanzania, a similar discovery of anatomically Modern Human remains, the Ngaloba skull, has been dated to between 150 and 120 thousand years ago.

In general, all of the archaic sapiens, the Neanderthals, Denisovans, and members of the ATG were larger, more robust, and more heavily built than we humans today, and Omo 1, Herto Man, and the Ngaloba sapien shared these attributes, even though their head shape and features were otherwise modern. But following their initial appearance, the average size of *Homo sapiens sapiens* individuals gradually diminished over the millennia until, by about fifty thousand years ago, it had reached the proportions of today's average human.

All members of the sapiens group, *Homo sapiens sapiens* included, inherited much the same technology from the erectines and appear to have continued to develop their technology and cultural practices along similar lines. Although widely separated geographically, the Neanderthals, the Denisovans, and members of the ATG all independently learned to make tools in the Mousterian tradition. Their toolkits included hafted hunting spears; stone and bone awls for sewing hides; finer, sharper scrapers for removing hair or flesh from hides; and sharp-edged blades for carving wood, bone, and ivory into more refined tools. All had acquired the technology of making fire on demand.

It remains unclear therefore, as to how and why fully anatomically modern Homo sapiens sapiens evolved in the Afar Triangle and, as far as is known to date, nowhere else. Whereas the Neanderthals appear to have evolved as lineal descendants of *Homo erectus heidelbergensis* at many places in Europe and the Denisovans from *Homo erectus pekinensis* in Asia, the descendancy of *Homo sapiens sapiens* is clouded by the fact that there were several sapiens subspecies that coexisted in Africa at the time of their appearance. Yet, it appears that only one group rather suddenly emerged with modern head anatomy. The scientific community remains divided between two possible theories, referred to as the *hybridization* and the *lineal descent models*, of how this happened.

The hybridization model hypothesizes that Homo sapiens sapiens was in fact a hybrid subspecies, the result of interbreeding amongst the various ATG subspecies that were dispersed across Africa between roughly four hundred and two hundred thousand years ago. In this model, it is proposed that different subspecies developed one or more different modern traits in terms of head shape and internal brain organization, and those traits were shared when the subspecies met and interbred, ultimately giving rise to today's human head shape and brain structure. Under this theory, the *Homo sapiens sapiens* subspecies, whose remains were found at Omo Kibish, represent a unique genetic assemblage of anatomical characteristics that came from multiple sapien subspecies and resulted in a single hybrid subspecies with the fully modern anatomical characteristics and cognitive capabilities. Supporters of the hybridization hypothesis point to the fact that three hundred thousand years ago, the earth was experiencing a warm, wet period, and North Africa, including the Sahara region, was green and lush, facilitating the migration of early human bands throughout the region. Hence, it is very likely that the various subspecies of the ATG would have met and socially interacted, producing hybrid offspring.

In the lineal descent model, *Homo sapiens sapiens* evolved directly from a *Homo erectus heidelbergensis* population that was living in the Afar region of Ethiopia, presumably between three hundred and two hundred thousand years ago. While other similar sapiens subspecies were evolving in parallel and had achieved nearly modern anatomy, only the one subspecies living in the Afar Triangle region of Ethiopia underwent the unique and fortuitous genetic mutations that ultimately resulted in fully modern head shape and restructured brain. This rewiring of the human brain could have happened suddenly, the result of one or two significant genetic mutations, or gradually through a series of favourable mutations. However it happened, those individuals with increasing intelligence would have had a significant survival advantage and consequently lived longer, producing more offspring. Over successive generations, the numbers of these cognitively advanced humans increased relatively quickly, and they emerged as the dominant sapien subspecies in the region.

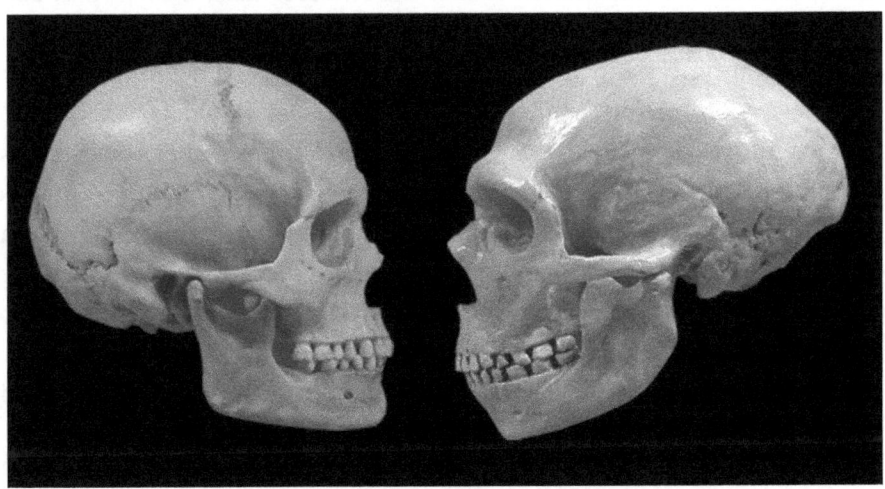

A comparison of a typical Modern Human skull (left) with that of a Neanderthal (right).

It may seem difficult to accept that a few changes in brain anatomy alone could account for such an astounding increase in survivability. Yet today, there are over seven billion *Homo sapiens sapiens* individuals living on Planet Earth and no Neanderthals or Denisovans, and the only apparent difference between them was brain anatomy and, by inference, cognitive capability. The intellectual capabilities of *Homo sapiens sapiens* must have been significantly more advanced for them to outcompete every other sapien group they encountered and to become so evolutionarily successful.

The differences in head shape between *Homo sapiens sapiens* and all other sapiens subspecies are subtle. They imply however, that the frontal and parietal lobes of the human brain are relatively larger and, by inference, more developed than in their sapiens cousins. Similarly, the lower rear portion, in particular the occipital lobe, is smaller and less developed in humans than in their sapiens cousins.

While the human brain is a highly integrated organ, different parts of the brain have been found to be associated with certain activities more than others. In this regard, the frontal and parietal lobes of the brain are primarily associated with cognition and language, while the mid and rear lobes are associated with visual information processing and motor coordination. The implication is that the Neanderthals, Denisovans, and other members of the ATG that had larger eyes and more advanced visual information processing very likely had superior visual acuity and better hand-eye coordination, while *Homo sapiens sapiens* would have had superior cognitive powers. The Neanderthals and other sapiens subspecies would have been more capable hunters in terms of slaying large animals in close quarter combat, while Homo sapiens sapiens would have had a greater capacity for abstract thought, language, strategizing, knowledge accumulation, and technology development, all of which allowed them to devise safer strategies to secure their meat.

With a more powerful intellect came superior social networking and organizational skills, enabling *Homo sapiens sapiens* to support and harmoniously maintain larger social groups. More advanced hunting and gathering strategies enabled them to provide the additional food and other resources required to support their growing populations. Specialization became possible as these early human societies could afford to have individuals spend more time doing what they did best, so everyday tasks could be assigned to different group members in accordance with their capabilities and skills. The ability to function in well-organized, larger social groups would have greatly enhanced *Homo sapiens sapiens*' survivability and given them an enormous competitive advantage over any other sapiens subspecies they encountered.

So it was that the elegant evolutionary principal of natural selection, operating in concert with the extreme environmental stresses of the Pleistocene, forged a species whose primary survival strategy was an extraordinary mind, a mind like nothing the earth had seen before. The cognitive capabilities of *Homo sapiens sapiens* were so superior that they enabled our ancient ancestors to ultimately outcompete every other animal on the planet and to not only survive as a species but to flourish in unprecedented numbers.

That is not to say, however, that *Homo sapiens sapiens*' survival was assured, at least not when they first appeared.

CHAPTER FIVE
Relying on Instinct

"It is impossible to overlook the extent to which civilization is built upon the renunciation of instinct."

...Sigmund Freud

PALEONTOLOGICAL EVIDENCE SHOWS that during the first several millennia of their existence, *Homo sapiens sapiens*' technology and lifestyle were not that different from the Neanderthals, Denisovans or other members of the African Transition Group. In spite of the latent intellectual capabilities that they possessed, more than one hundred thousand years would pass before their cognitive powers gave rise to the cultural explosion that is the hallmark of humankind today. During those first one hundred thousand years, our ancient ancestors relied on the technologies, social practices, and instinctive behaviours that they had inherited from their erectine ancestors.

When *Homo sapiens sapiens* first appeared in the Great Rift Valley of east Africa, they would have seemed rather poorly equipped for survival compared with most of the other animals on the savanna. They had no large canine teeth or extendable claws with which to kill prey or defend themselves. They were much slower runners than the prey they hunted and the predators that hunted them. These first humans were relatively large creatures and could not readily hide. Unlike other primates, humans had weak forearms and were unable to forage in the safety of forest canopies or to climb trees fast enough to escape predators. Their eyesight, hearing, and sense of smell were average at best, and they were hairless, making them vulnerable to climate change. In spite of their inherent intellectual capabilities, a lone individual would have been easy prey for the lions, cheetahs, and other swift predators that hunted on the savanna at the time.

Homo sapiens sapiens' young were born completely helpless and remained totally dependent on their mothers for food and protection for several years following birth. Mothers, in turn, were dependent on other band members for protection and assistance in rearing their young. If such support was not forthcoming, because of separation, ostracism, or inability of the band to support additional members for any number of reasons, both mother and infant would have most certainly perished.

To make survival even more challenging, between 200 and 130 thousand years ago, the earth was in the final stages of the Pleistocene epoch, and its climate was growing ever cooler and drier as the last and most severe ice age advanced from the poles, inexorably extending its icy grip towards the temperate and tropical regions of the earth. From the time the sapiens appeared, about four hundred thousand years ago, to the height of the last great ice age, about eighteen thousand years ago, the earth's annual temperature averaged some 2 to 3°C colder than today.

But the climate was also highly cyclical, having prolonged cold periods when the temperatures plunged to 4 to 5°C colder that were punctuated with short warm periods when the earth's temperature rose to 1 or 2°C warmer than it is today. During the brief warm periods, the savanna greened and became a nourishing grassland that was interspersed with woodlands. Tropical rain forests expanded, lakes and rivers filled, and marshes formed, providing wetland ecosystems for a vast diversity of flora and fauna. During some of these warm periods, even the Sahara region of North Africa was turned into a lush, green ecosystem of grasslands, woodlands, and flowing rivers. But during the cold periods, the east African savanna became cold and dry, grasslands withered and died, and the lakes and rivers of the Great Rift Valley were much diminished or completely dry.

Animals living on the savanna, including *Homo sapiens sapiens*, had to adapt to such changing conditions or perish. Sometimes conditions were so severe that all life forms had to leave areas of the savanna in search of more hospitable places to live. At other times, conditions were much like today, and the flora and fauna of the savanna flourished in the warmer temperatures and regular, seasonal rainfall.

Two hundred thousand years ago, *Homo sapiens sapiens'* survival as a species was by no means a certainty. The challenges imposed by the cyclical climate and changing ecosystems tested their survival skills to their very limits, and during times of extreme cold and minimal precipitation, bands would have perished.

The paleontological evidence indicates that the Pleistocene glacial events took a heavy toll on human life, and during at least one such event that occurred around 170 thousand years ago, *Homo sapiens sapiens* came perilously close to extinction. Bands of perhaps one to two dozen souls became widely dispersed, and some studies have estimated that the total human population may have been reduced to as few as two or three thousand individuals. However, the evolutionary processes of natural selection had shaped our earliest ancestors into the world's ultimate social animal, providing them with a powerful, instinctive capacity to work together to overcome almost any adversity the vagaries of the Pleistocene wreaked upon them.

Instinct as a Survival Strategy

Instinctive behaviour is the fundamental survival mechanism of all living organisms. All life forms, from bacteria to humans, rely to some degree on instinct to help them perform the basic activities required to survive—to find sustenance, escape predators, and to reproduce. Instinctive behaviour is innate, encoded in an organism's DNA. It is something with which all animals are born (or hatched), and it plays a larger role in governing the behaviour of lower life forms than higher ones. All insect behaviour is instinctive, and most mammal behaviour is learned, but all animals, humans included, rely on instinct to a greater or lesser degree.

Instinctive behaviour is responsible for some of the most remarkable behaviour in the animal kingdom. Newly hatched sea turtles for example, automatically scurry over the beach towards the ocean as soon as they dig their way through the sand covering the nest. While they can probably smell the ocean to determine the direction to go, how do they know to associate the smell with water or that they are intended to spend their lives in the ocean? They have never experienced water and have no parents to teach or guide them as their mother has long gone back to the sea. And how do they know, after some twenty years of foraging over twenty thousand kilometres across the major oceans of the world, to return to the exact same beach where they hatched? In fact, they don't *know* in the sense of conscious thought; they are acting instinctively, automatically following behavioural patterns that have been encoded in their DNA over the millennia.

Animal migration also stems from an instinctive impulse to change habitat or return to a place of birth, typically triggered by subtle, often seasonal, changes in an animal's environment. Whales, salmon, caribou, many species of insects, and, of course, almost all species of birds in the Northern Hemisphere follow migration patterns driven by instinct, and some migration behaviour is truly astounding.

One astonishing example of instinctual migration involves the Monarch butterfly. Monarch butterflies spend most of their lives migrating over four thousand kilometres from wintering grounds in southern California and Mexico to the northern United States and southern Canada for the summer. This behaviour is impressive, but when you realize that the lifespan of an adult is measured in weeks, and it takes four generations of them to complete a single migration round-trip, the complexity of the Monarch's instinctive behaviour is truly astonishing.

Instinct also provides the foundation for social behaviour. Many animals, from insect colonies to wolf packs, Meerkat clans, and human societies, depend on organized social behaviour for survival, and in the insect world, this instinctive social organization is truly remarkable.

In most insect colonies, the population does not consist of identical individuals. Instead, groups of individuals within a colony have evolved with different specialized capabilities. In a honeybee colony, for example, each individual carries out the specific tasks that they evolved to perform; each is dependent on the others in order to survive, and the survival of the entire colony is dependent on each individual performing the role they evolved to fulfill.

The queen bee is supreme. Only she can lay fertilized eggs, and a typical queen can produce up to two thousand eggs within a single day. A queen mates early in life, and millions of sperm are retained within her body so she can continue to populate the hive for three to five years. Her unfortunate mates, the drones, are the only males in the colony, and their single task is to fertilize a new queen. Drones mate with a newly hatched queen outdoors, usually in flight, and die soon afterwards. Any drones still alive after the fertilized queen returns to the hive are vanquished or sometimes killed by the worker bees, the largest population within a colony. Worker bees are exclusively female, but they cannot produce fertilized eggs. They forage for pollen and nectar, tend to the queen and drones, feed larvae, ventilate the hive, defend the nest, and perform other tasks to ensure the survival of the colony. It comes as no surprise that the average lifespan of a worker bee is only six weeks as they literally work themselves to death. In the honeybee hive and other such colonies, the evolutionary principle of survival of the fittest takes on a whole new dimension as it no longer applies to individuals, but rather to the social organization of the colony as a whole.

However impressive and even inspirational as some of nature's instinctive behaviour is, its inflexibility makes it a very fragile and vulnerable survival mechanism. In most animals, instinctive behaviour is hard-wired, meaning that if conditions change so that a certain instinctive behavioural trait is no longer beneficial or is detrimental, the animals lack the intellectual capacity to consciously change or adapt. Animals that rely heavily on instinctive behaviour cannot adapt to sudden changes in environmental conditions, and they often perish when such changes occur. For example, along the Monarch butterfly's migration routes, much of the traditional habitat has been reduced or destroyed, and the population of these beautiful insects has plummeted as a result. Sadly, they are unable to comprehend the need to change their habits by altering their migration routes or sampling different foods.

While early humans would have initially relied to a great extent on instinct, over time they learned to consciously override or intellectualize much of their instinctive behaviour, and today most adult human behaviour is learned. Even in prehistoric times, much of what other animals did instinctively, *Homo sapiens sapiens* would have gradually intellectualized and used to advantage. Intellectualizing their behaviour enabled early humans to become more adaptable to change and increased their chances of survival under a wide variety of environmental conditions. It also allowed them to recognize and anticipate instinctive behaviour in other animals and to respond by modifying their hunting practices, increasing their success at killing prey with less risk to themselves.

There are, however, vestiges of instinctive behaviour in we humans even today that prompt us to occasionally do things impulsively. A simple example is the startle reaction whereby we react defensively when surprised by a sudden sound or unidentified movement seen out of the corner of our eye. Some of us involuntarily recoil at the sudden sight of a snake, an insect, or a rodent. Such behaviour is more likely than not an ancient survival mechanism that has been subtly encoded into our genetic fabric over the eons, and amusing though it may be in today's world, it might very well stem from a life-saving instinctive impulse for an early human on the African savanna.

Our reliance on instinct also evolves during the course of our lifetimes. Babies are born with instincts that ensure they get a healthy and safe start in life. The suckling instinct, for example, develops while a baby is still in the womb, as evidenced by ultrasound images of unborn babies sucking their thumb. Newborn babies instinctively suckle at a mother's breast immediately after birth and cry out to attract attention when in need of food, protection, or comfort. As the infant develops into a child, such behaviours become "intellectualized" as the child learns to address its own needs for food, comfort, and safety.

For the first several years of life, human behaviour is primarily self-centred, an evolutionary adaptation that helps ensure the survival of the individual. Gradually, however, as a person matures into adulthood, other more subtle instinctive impulses emerge that reinforce behaviour that is beneficial to survival of the species as a whole. In particular, many of our emotions and emotional responses, as manifest in our conscience, are instinctive, and they guide us to act and behave in ways that maintain harmony with friends and family, and such socially beneficial behaviours contribute to the survival of our larger social group and community.

In prehistoric times, as well as modern, humans have relied on one another to survive and flourish, cooperating in the hunting and harvesting of food and other life-sustaining resources; in the development of technological tools, devices and weapons; and in the defence of the community. We help one another during times of illness, injury, or the infirmity that comes with old age. Such social behaviour is guided by our emotions, and emotional bonding with one another, which is, in turn, rooted in instinct, bearing traces to the time of the Australopithecines, and perhaps even before that. Social organization and cooperation played a major role in hominin survival in ancient times and remains critically important today. Very few individual humans can survive without the support of others, not two hundred thousand years ago and not in modern times.

The most fundamental behavioural trait that is observed in most social animal colonies is the *priority of group well-being over that of the individual*. In most social groups, individual behaviour has evolved to share awareness of impending threats or location of food and other resources. Individuals invariably cooperate in the life-sustaining, daily activities of the group, and in many cases, group members willingly sacrifice their own desires and needs, even their lives, to ensure the safety of the group.

While some animal social groups are dominated by a single individual or dominant pair, most social animal groups are *egalitarian*, having no individual deemed more worthy than any other. True, individuals within social groups assume different roles, and within some of those roles, such as leadership, individuals enjoy the privilege of position and may use it to force other members to conform to certain behaviour. However, the enforced behaviour is usually necessary for the well-being of the group, and roles such as leadership are still carried out within an egalitarian framework. Egalitarianism remains the underlying principle of democratic governments and institutions throughout the world in modern times.

All social animal groups have an instinctive behavioural pattern within which every individual is expected to conduct themselves, and the expectation is generally the same for all members of the group. This *code of behaviour*, as it were, typically ensures the survival of the group and helps with the maintenance of harmonious and healthy interactions and relationships amongst the individual members. Failure to comply with the tenets of this behavioural framework typically results in group members taking corrective action or expelling the offender from the group.

Human emotion plays a major role in social bonding and ensuring harmonious behaviour with social groups. We humans experience a vast array of emotions, and the intensity of our emotions varies in response to the intensity of the experience eliciting them. Throughout the ages, many psychologists have wrestled with understanding the nature of emotion and the role it plays in the conduct of our daily lives. Most suggest that we have a relatively small number of basic or primary emotions—estimates vary from as few as six to as many as twenty-seven—from which all other emotions are either variations in intensity or the result of combinations of the basic emotions. This concept dates back to the first century BCE when the ancient Chinese *Book of Rites* identified the seven basic *feelings of men* as joy, anger, sadness, fear, love, disliking, and liking.

In more recent times, the psychologist Robert Plutchik published a classification scheme for general emotional responses wherein he proposed eight primary emotions, grouped into four pairs of polar opposites: joy-sadness, anger-fear, trust-distrust, surprise-anticipation. Other emotions, he argued, represent either variations in intensity of the primary emotions or combinations of them.

Plutchik represented the entire range of emotions in a *Wheel of Emotions* diagram, depicting the eight primary emotions as polar-opposites. Plutchik's diagram illustrates how similar emotions represent different levels of intensity of a primary emotion; the more subtle or complex secondary emotions relate to one another as combinations of the primary ones. In Plutchik's diagram, the intensity of emotion and the shading associated with it, increase as you move toward the wheel's centre.

Plutchik also published a *Psycho-evolutionary Theory of Basic Emotions* wherein he proposed that all animals experience emotion, and those emotions have primeval origins, moulded by evolution over millions of years. He further postulated that emotions better enable all animals to adapt to changes in their environment, to varying degrees, thereby ensuring the survival of their species.

Plutchik postulated that the basic human emotions evolved in response to the ecological challenges faced by our ancient ancestors and are so primitive as to be genetically *hard-wired*, with each basic emotion corresponding to a "dedicated neurological circuit within the brain." In other words, during the course of evolution, our basic emotions were genetically encoded in our DNA, and our emotional reactions are instinctive. Like all instinctive behaviour, our basic emotions and emotional reactions are innate and automatic, triggering behaviour without thought and often without the ability to control them.

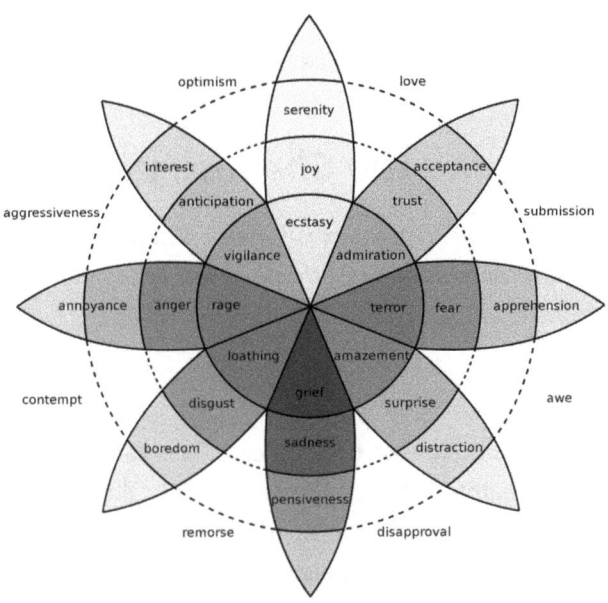

Robert Plutchik's *Wheel of Emotions*.

Plutchik argued for the primacy of the eight basic emotions by showing that they have the highest survival value. The *apprehension, fear, and terror emotion* triggers the fight or flight response, helping to ensure that an individual makes the appropriate decision in terms of their survival. The *acceptance, trust, and admiration emotion* leads to social compatibility, intimacy, and reproduction, strengthening bonding within social groups and ensuring survival of the species.

Plutchik's model of polar-opposites is also revealing in terms of how emotions provide us with the ability to fully appreciate the richness of the life experience. For until one has experienced pain, one cannot fully appreciate the true enjoyment of pleasure, or without having experienced rejection or indifference, one cannot fully appreciate the ecstasy that comes with romantic love.

As Plutchik has argued, our emotions have played, and continue to play, a major role in our survival as a species. But our emotions do so much more than merely helping us to survive as an individual and a species; they are also what enable us to appreciate the beauty and wisdom in the natural world and to fully embrace the experience of living. They are the foundation of our culture and add untold richness to our lives.

Evolution and the Chemistry of Emotion

> *"Love is Chemistry. Sex is Physics. But Marriage is Quantum Mechanics and I think it's fair to say that no one truly understands Quantum Mechanics."*
>
> *...Anonymous*

Since *Homo sapiens sapiens* first appeared on earth, their survival has depended on the maintenance of the family unit, strong social bonding, and cooperation. Over the course of human evolution, natural selective stresses forged the body's physiological systems to reinforce those types of behaviour that would promote life partnering and social bonding. These behaviours, in turn, ensure the successful rearing of children, the maintenance of harmony within our social group, and ultimately, the survival of our species.

The body's primary regulator of emotional state is the *endocrine system.*

The endocrine system is a collection of glands that secrete hormones, neurotransmitters, and other biochemicals directly into the circulatory system to be carried to the brain and other organs throughout the body. Within the brain, these biochemicals alter the perceptions, thoughts, and emotions that constitute our mental and emotional state.

The endocrine system is responsible for our conscious feelings and emotional responses to the wide range of stimulation received from our five senses, providing the mechanism by which we react to events around us and experience being alive. These sensory stimulations, referred to as *stress factors* or *stressors*, range from perceived physical threats, which generate feelings of fear or anger and trigger the fight or flight response, to the sound of a beautiful piece of music, which can elicit feelings of contentment, well-being, happiness, and joy. Many chemical reactions take place in the brain at any given time, constantly regulating a myriad of bodily functions, preparing the body's response to external stressors and, in the process, generating emotional responses in our consciousness.

Within the endocrine system, the hormones and neurotransmitters serotonin, oxytocin, dopamine, and endorphins are the primary biochemicals that alter our emotional state. They are sometimes referred to as the *happy chemicals* and are largely responsible for triggering pleasant feelings and maintaining a sense of well-being. In response to generally positive sensory perceptions, our system releases increased amounts of these biochemicals into our brains, and we experience agreeable feelings in proportion to the amount of chemical released. This positive response can range from contentment to happiness, joy, and euphoria. Conversely, in response to negative perceptions or experiences, the supply of these hormones is diminished, and we feel disturbed, angry, sad, depressed, and ultimately suicidal.

An increased level of serotonin elevates our mood and makes us more agreeable and sociable. It enhances our self-esteem and promotes a sense of well-being and happiness. It reinforces positive relationships with family, friends, and members of our social group. Serotonin flows when we feel valued and appreciated for what we contribute to our family or community. It is serotonin that makes us feel good when we help others and are respected and appreciated in return. Conversely, low levels of serotonin can cause depression, anxiety, and a loss of self-esteem, particularly when associated with problems related to social interaction or fear of rejection by friends, family, or peers. It is primarily serotonin that motivates people to treat one another with love and respect, the emotional fabric that enables human societies to grow into millions of people while maintaining peace and harmony.

Oxytocin promotes emotional attachment, fidelity, and the nurturing of healthy relationships with one's life partner, family members, and friends. It is sometimes called the love hormone because it is responsible for engendering feelings of love, infatuation, trust, and attachment. Oxytocin is released by both men and women during intimacy, by mothers during breastfeeding, and when friends or family members embrace, hold hands, or touch. Simple gestures of affection, such as hugging and gift-giving, raise oxytocin levels. Oxytocin is essential for creating strong interpersonal bonds and improved social interactions, and low levels result in irritability and an inability to feel affection or form close bonds with others, often leaving one feeling that life has no joy. It is primarily oxytocin that produces the strong emotional bonding that is the foundation of the *nuclear family*.

Dopamine has the effect of strengthening our will to live and motivates us to take action toward achieving goals or acting on our desires and needs, giving us that surge of pleasure or a sense of achievement when we accomplish a goal. Imagine the dopamine levels that must course through the veins of Olympic gold medallists as they stand on the podium listening to their country's national anthem and watching their flag being raised. A strong drive to achieve or win is the result of high levels of dopamine, while procrastination, self-doubt, and lack of enthusiasm are linked with low levels.

Dopamine is also released when we have even the simplest of pleasurable experiences, such as holding our newborn child for the first time, listening to a favourite piece of music, watching a sunset, or taking a walk in the forest. The more enjoyable we find the experience, the greater the flow of dopamine into our brain. During particularly enjoyable experiences, dopamine levels surge, and we can have an exceptionally intense emotional response that brings tears to our eyes and fills us with such joy that it feels like our hearts will burst with the intensity. Such intense emotional responses make us happy to be alive and are what make the life experience so special.

Endorphins are released in a number of circumstances. For instance, in response to pain and stress, they help us cope, acting as both analgesic and sedative and diminishing the perception of pain. They trigger a surge of energy during intense physical exercise, and they are responsible for both the second wind and runner's high that marathon runners experience. Endorphins can also be flooded into the system with dopamine, resulting in an intensely orgasmic feeling—euphoria and joy—experienced when pain and pleasure occur simultaneously at the marathoner's finish line or at sexual climax.

Through the release of hormones and the pleasant feelings they produce, the endocrine system rewards us for certain behaviours and punishes us for others. As a result, we learn both consciously and unconsciously to associate pleasant feelings with certain behaviours and develop a form of addiction to endocrine chemicals in that we learn to routinely behave in specific ways so we can experience the enjoyable feelings that they produce.

In evolutionary terms, dopamine, serotonin, and oxytocin would have reinforced the formation of the family unit and close-knit social groups amongst early humans by rewarding socially beneficial behaviour. Such behaviour would have been accompanied by feelings of love, friendship, contentment, and happiness, and the strong sense of self-worth that comes with having a life partner and a family, and being a valued, contributing member of one's community. In this way, the endocrine system guided early humans towards the development of strong bonds with their families and friends and to actively contribute to the well-being of their societies. The maintenance of the family unit and a harmonious social organization have always been, and continue to be, essential to human survival.

The Cohesive Bonds of Kinship and Humanity

"I will never apologize for saying that the future of humanity and the future of the world is going to be defined by what we have in common as opposed to those things that separate us and ultimately lead us into conflict."

...Barack Obama

The organizational cornerstone of almost every human group, whether it is a prehistoric band of a few individuals or a modern society of millions of people, is the *nuclear family*. A nuclear family is typically based on a monogamous relationship between a male and a female who stay and work together in the caring for their offspring into adulthood. The emotional bonding within a nuclear family is extremely strong, providing its members with an enduring motivation to care for and protect one another throughout their lifetimes. In addition, multiple nuclear families living in close proximity typically share important values and are motivated to work together to create larger, harmonious communities wherein individuals can enjoy life with friends, family, and members of their community at large. Within such communities, more complex social organizations can be developed, and the communities can grow into cities of millions of people where its citizens prosper and culture blossoms.

While there are a large number of social animals throughout the animal kingdom, the nuclear family–centric organization is unique to humans. Many mammalian social groups are led by a dominant alpha pair, and only the alpha pair are allowed to breed. Our primate relatives, monkeys, chimpanzees, and gorillas, live in groups organized as either a *one-male–several-female group* wherein only the dominant male is allowed to mate with multiple females, or in *multi-male–multi-female groups* wherein both males and females have a number of different mates. Societies organized along either of these lines tend to be limited in size. Gorilla troops, typically organized as a dominant male with several females and their young, rarely grow to more than thirty individuals. Chimpanzee troops are typically organized as *multi-male–multi-female* communities, but when the population exceeds about 100 to 150 individuals, groups become unstable and usually divide into small bands that go their separate ways.

It cannot be known with any degree of certainty when nuclear family bonding first appeared, but the discovery of the *First Family* remains in the Awash Valley of Ethiopia suggests that it was likely already present in the Australopithecines and was no doubt the cornerstone of bands of erectines and other species of hominins that followed them. The first bands of *Homo sapiens sapiens* to roam the African savanna would have been extended family groups that varied in size from a few to several dozen individuals. These extended family groups were held together by a strong cohesive bond referred to today as *kinship*, which came about through a combination of emotional attachment, blood ties, and the sense of belonging that comes with being a contributing member of a social community.

The sense of self-esteem that comes from feeling valued by fellow members of one's social group is perhaps the strongest element underlying human social bonding and kinship. It causes the individual's self-perception to be inseparable from the group and creates powerful bonding at a deep emotional and psychological level. Today, humans are not much different. Our identity and sense of self-worth are often linked to our work, our contribution to society, or both.

In today's global society of over seven billion people, the cohesive bonds of kinship are no longer a factor in maintaining peace and harmony throughout the world. But vestiges of the social impulses that underlay kinship bonding are still manifested in our conscience and what is referred to as a *sense of humanity,* those innate feelings of love for our fellow human beings that prompt us to treat one another with benevolence, dignity, and respect.

In our modern world, our humanity provides the cohesive bonding that maintains our social order across the globe, and our societies rely on this humanity to function and flourish. It is our humanity that enables us to develop and enjoy our cultures; to maintain peace, a spirit of cooperation, and mutual respect throughout the world; and to provide a harmonious social environment wherein we, as individuals, can enjoy our lives with family, friends, and members of our community.

Today, however, religious and ideological conflicts are ravaging societies in many parts of the world, and we are witnessing the apparent loss of humanity and collapse of social order. We see societies being subjugated by despots who destroy cultures and force their people to flee to neighbouring countries for fear of their lives. Even in a few larger, so-called developed nations, we see the rise of leaders that act more in their own self-interest than in the best interests of their citizens, undermining the moral and ethical fabric that holds such nations together.

At such times, it seems that humankind is in danger of losing the cohesive bonds our humanity affords and that the world social order is on the brink of descending into anarchy. But even in the darkest of such times, when faced with a larger, more global threat, such as the coronavirus pandemic of 2020, our humanity re-emerges and we pull together, helping one another in our defence against a common adversary.

It is only our humanity, it seems, that enables humankind to come together to defend themselves against perils that pose a threat to their existence. Today, humankind is facing ever-mounting threats in the form of climate change, environmental collapse, overpopulation, and moral and ethical decay. Let us hope that our humanity will soon re-emerge and enable us to put our differences aside and pull together to address these looming threats to our species, indeed to all living species on Planet Earth, before it is too late.

There is much wisdom inherent in our instincts, forged by the processes of evolution over millions of years, and it behooves we humans to heed them, particularly with regard to our emotions, our conscience, and our sense of humanity. It is these instinctive impulses that have provided the foundation for our astounding evolutionary success to date and will guide us through the threats to our survival that are now most certainly upon us.

Our emotions, and the physiological systems that create them, are a marvel of biological adaptation and yet another testimony to the power and wisdom inherent in the simple, yet elegant, processes of evolution by natural selection.

CHAPTER SIX

The Great Rift Valley – Land of Our Birth

"A man's homeland is wherever he prospers."

...Aristophanes

THE GREAT RIFT Valley is a massive, serpentine trench that stretches from Lebanon in the Middle East, across the Red Sea, and south through the east African countries of Eritrea, Ethiopia, Kenya, Uganda, Tanzania, and Malawi to Mozambique. The valley is the result of the divergence of tectonic plates that began deep under the earth some thirty-five million years ago when the Arabian tectonic plate began to drift eastward, separating from the African plate. As the plates diverged, enormous lateral forces were generated, ripping the earth's crust apart and causing large sections of the land to sink between parallel fault lines, a geological process called *rifting*.

Over millions of years, as the rift widened, the valley floor further subsided, and today, some sections lie well below sea level. Near its northern end, the Dead Sea is the lowest land-based feature on earth, its surface lying some 430 metres below sea level. In its central portion, the valley separates into the Eastern or Albertine and the Western or Gregory rifts, and lying between the rifts are the African Great Lakes—Turkana,

Tanganyika, and Victoria—some of which are the deepest on the planet. Further south, the Eastern and Western rifts join at Lake Malawi. From there, the rift continues south along the Shire and Zambezi river systems, ultimately reaching the shores of the Indian Ocean in Mozambique. The valley floor continues to subside, and geologists tell us that in some tens of million years hence, waters of the Red Sea and the Indian Ocean will breach the highlands, flooding the valley and separating east Africa from the rest of the continent.

Long ago, the Great Rift Valley provided early hominins with access to neighbouring regions of the African continent. Paleontological evidence of ancient human occupation has been found in coastal cave complexes along South Africa's coastline, the Kalahari region of Botswana, and the large complex of limestone caves known as the Cradle of Humankind, located in the Veld region of South Africa. The Cradle of Humankind cave system, in particular, has proven to be one of the world's richest sites of ancient hominin remains and has been declared a UNESCO World Heritage site. Discoveries made there date back four million years and include the remains of the *Australopithecines africanus* and *sediba*, and *Homo ergaster*. It was in the nearby Wonderwerk Cave that *Homo ergaster* built the world's first firepit and mastered the art of controlling fire for their own use, some one million years ago.

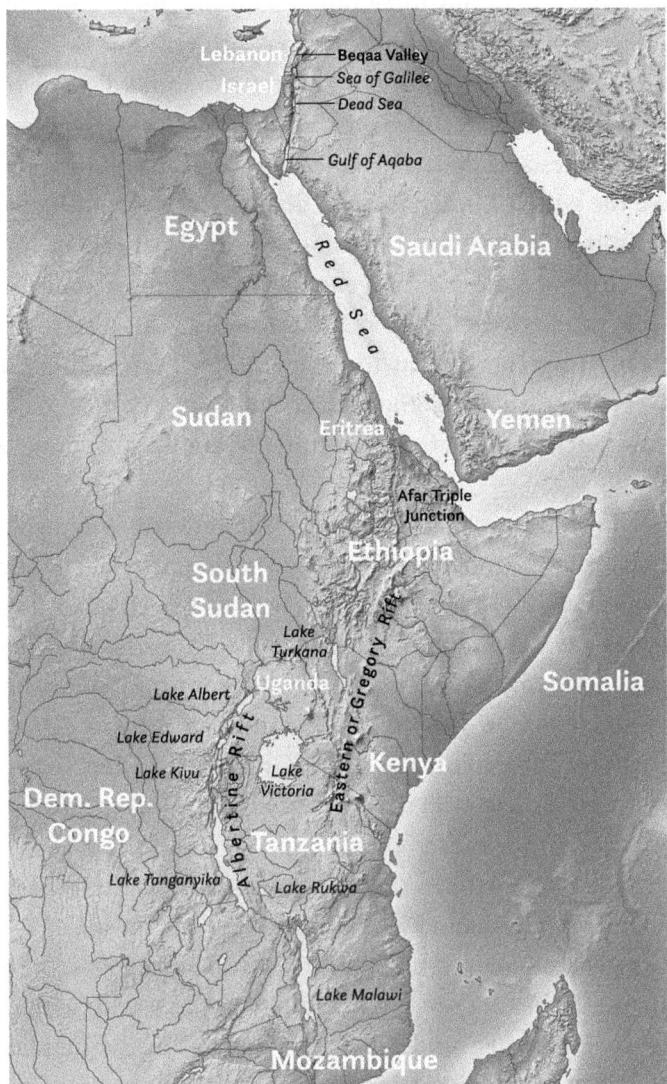

A map of east Africa showing the Great Rift Valley, the Afar Triangle, and the Nile River system.

Paleontological evidence indicates that for the first few millennia or so after their arrival, *Homo sapiens sapiens* remained largely in the vicinity of the northern Great Rift Valley. But around 170 thousand years ago, what appears to have been an extreme climate event significantly reduced the human population in the area to as few as a few thousand individuals.

The harsh living conditions forced the remaining inhabitants to leave the region and to disperse across the African continent.

Africa today is home to more than three thousand distinct tribes, with different physical characteristics and culture, speaking over two thousand different languages and dialects. Genetic research has shown that African tribes have the widest human genetic diversity of any continent on earth, stemming from the separation of tribes for long periods of time in geographically diverse regions. The genetic evidence also confirms that this separation began between 200 and 170 thousand years ago.

These early migrants appear to have left the savanna in all directions, although evidence suggests that from the Afar Triangle region, much of this intra-African dispersal was initially southward along the Great Rift Valley.

While today there are many distinctly different tribes living throughout the valley, speaking over three hundred different dialects and practising very different cultures, a recent analysis of the DNA of these eastern African tribes has shown less diversity than tribes living elsewhere around the African continent, indicating that in ancient times, they were less isolated from one another. Due to the openness of the east African terrain and the ease of movement along the Great Rift Valley, it appears that these ancient bands and tribes moved frequently and socialized with one another whenever they met.

In general, the east African tribes are unusually tall, perhaps reflecting a heritage of life on the savanna where longer legs and an ability to run faster and over longer distances were distinct survival advantages. In this regard, some Sudanese tribes are among the tallest in the world, with an average adult male height of just under two metres, and in the highlands of the Great Rift Valley of neighbouring Kenya, the Kalenjin tribe is renowned for its disproportionate share of long-distance running champions. Since 1988, twenty of the twenty-five first-place male finishers in the Boston Marathon have been Kenyan, and while Kenya has a population of over forty million, three-quarters of these marathon champions are from the Kalenjin tribe with a population of less than five million.

As early *Homo sapiens sapiens* bands were dispersing southward along the valley, they eventually made their way along the Shire and Zambesi river systems to the coast of the Indian Ocean. Evidence at the Pinnacle Point Caves further south along the coast of South Africa shows they had reached that site as early as 170 thousand years ago.

In sharp contrast to the tribes living in and around the Great Rift Valley, a genetic analysis of the DNA of the Khoisan San people, who live around the Makgadikgadi Basin of northern Botswana, has revealed that they are genetically divergent from all other humans and are possibly the most ancient of human lineages living on earth today. Such evidence suggests that shortly after their first appearance some two hundred thousand years ago, bands of *Homo sapiens sapiens* migrated southward along the Great Rift Valley and across into the Kalahari region of Botswana, having little contact with any other tribes.

So ancient is the Khoisan San DNA that some archaeologists argue that the genetic findings indicate that Homo sapiens sapiens may, in fact, have first appeared in the Makgadikgadi Basin rather than in the Afar Triangle of Ethiopia. While no paleontological remains have been found to support the claim, the age of the Khoisan San DNA is not in question. Given their ancient origins, the people of the Kalahari, known as San Bushmen, may provide insight into the lifestyle of our most ancient ancestors.

About three thousand San Bushmen still follow a traditional lifestyle of hunting and gathering wild food. They live in social units of ten to fifteen individuals, and their possessions consist of only what they can carry. The only concession most have made to the modern era is that they wear clothing and have adopted some basic food staples.

Hunting large game on the desert and grasslands is exciting and dangerous, yet the San Bushmen still hunt enthusiastically in accordance with their ancient tradition. The San hunters still practice persistence hunting, and their techniques have changed very little since prehistoric times. After locating and isolating an antelope or kudu from a herd, they begin to pursue it across the desert and grasslands in the midday heat when temperatures typically reach 40 to 42°C. The chase typically lasts for many hours, sometimes even days, and covers a distance of several kilometres. During the chase, the pursued ungulate often runs out of sight and stops to rest in a shady area, but the persistent hunters track it down before it has had enough time to catch its breath. The animal is repeatedly chased and tracked down in this manner until it is too exhausted to continue running, and the San hunters can approach close enough to kill it with their spears. When the men return from the hunt, all gather around a cooking fire, and the women sing and beat various musical patterns with handclapping and drums, while the men dance in celebration of their victories.

The most prized game is the eland, the largest antelope in southern Africa. The eland has been a part of the San culture for millennia. Ancient San nomads practiced a simple form of spirituality, expressed in their songs, stories, and particularly art. They identified spiritually with animals and practiced shamanism whereby they transport their spirit to the spirit world. There, they communed with the spirit of animals, most particularly the eland. Over the millennia, the eland has become embedded in the San Bushman culture, and they continue to believe they have a close, spiritual relationship with it.

In addition to dispersing southward along the Great Rift Valley, some tribes of early humans dispersed northwest from the Afar Triangle, eventually reaching the Nile River valley.

About 120 thousand years ago, the earth was in the midst of an interglacial and the Nile River valley, along with the neighbouring Sahara and Levant regions, were green and lush, with grasslands, forests, lakes, and rivers. Paleontological evidence shows that it was about this time that *Homo sapiens sapiens* first arrived in the Sahara region, eventually reaching as far west as present-day Morocco.

Perhaps migrating south from the Sahara region, or north from the Kalahari, early humans eventually reached the Congo Basin by about one hundred thousand years ago. At the time, the Congo region was much like it is today: a vast, dense, tropical rain forest covering over four million square kilometres.

The Congo is the present-day habitat of the Efé pygmy people, a hunter-gatherer society living in the depths of the rain forest. A few Efé maintain a lifestyle much the same as their ancient ancestors, constructing simple leaf huts for shelter and living off the natural bounty of the rain forest and its many rivers. They hunt with bow and arrow and fish with bone-tipped, barbed harpoons that are fashioned much the same way as the Semliki harpoons discovered at an archaeological site in the region and dated to ninety thousand years ago.

The Efé are among the shortest peoples in the world. An average male's height is 142 centimetres. They have been found, through genetic analysis, to have lived more or less in isolation for over ninety thousand years. A diminutive stature would have proven an advantage when hunting in a dense rain forest, and it is likely that such environmental factors contributed to the gradual decrease in height of the Efé over the millennia.

As these early human hunter-gatherer bands wandered across the African continent, they encountered vastly different environments, including grasslands and woodlands, coastal plains and shorelines, hot desert sands, high mountainous regions with forested valleys, and the hot, humid jungles of the Congo Basin. These different ecosystems had unfamiliar plants and animals, prey and predator alike, and bands of *Homo sapiens sapiens* had to learn new hunting strategies, develop new lifestyles, and acquire new knowledge of which plants and animals were safe to eat and which were not.

For most species of animals, adapting to such extreme environmental changes would prove virtually impossible, and most would perish. *Homo sapiens sapiens*, however, proved to be remarkably resourceful and adaptable, so they not only survived, they thrived. Over the millennia, band sizes gradually expanded into tribes of a hundred or so individuals and by about seventy thousand years ago, *Homo sapiens sapiens* could be found in virtually every habitable region of the African continent.

CHAPTER SEVEN
The Dawn of Human Spirituality and Culture

"Every man lives in two realms: the internal and the external. The internal is that realm of spiritual ends expressed in art, literature, morals, and religion. The external is that complex of devices, techniques, mechanisms, and instrumentalities by means of which we live."

...Martin Luther King, Jr.

HUMAN CULTURE HAS both utilitarian and aesthetic elements, which spring from our scientific knowledge and spiritual beliefs, respectively. Scientific knowledge is the product of our intellect, and it enhances our survival both as individuals and as a species. Spiritual knowledge springs largely from our emotions, which are also the means by which we enjoy the experience of living. Together, they harmonize in our consciousness, enriching our lives and enabling us to embrace and live our lives to the fullest.

A society's culture develops as an integrated whole, with the principal elements of knowledge, technology, language, customs, art, and spirituality all advancing in concert, with each element influenced by the others.

Throughout recorded history, scientific knowledge has been strongly influenced by spiritual or religious belief, and at least until relatively modern times, religious belief, in turn, has embraced scientific discoveries as they occur.

Our *scientific knowledge* fuels our creativity and is the basis of our technological development. The fabrication of tools, machines, and devices are a major component of our culture, enhancing our daily lives and enabling us to more easily acquire the essential elements we need to survive and flourish. Our technological inventions extend our lifetimes and remove much of the tedium in our daily routine by automating or making repetitive tasks easier. They facilitate better communication, entertain us, and help us feel more comfortable and secure.

Our *spirituality* fuels our emotions, enabling us to appreciate the beauty to be found in nature and to contemplate the spiritual or divine aspects of our existence. In this regard, our spirituality can be thought of as having both an *aesthetic* and a *divine* aspect.

It is through our *aesthetic spirituality* that we recognize and appreciate the beauty in the natural world—the delicate fragrance of a hyacinth borne on a summer breeze or the hummingbird's iridescent colours shimmering in the sunlight. And it is through our *divine spirituality* that we are aware of the spiritual aspects of our being, contemplate the awesome workings of the cosmos, and wonder about the nature of its creator.

Consider a honeybee foraging for nectar in a cherry blossom. Using our scientific knowledge, our logical, scientific mind turns to thoughts of how the bee collects nectar to take back to its hive where it will use the nectar to make honey. In the process, the bee has pollinated the cherry blossom, so in a few weeks' time, it will yield a cherry that we, or a bird, will enjoy.

Our aesthetic spirituality elicits an emotional response in our soul, whereby we appreciate and enjoy the sight, smell, and sound of the industrious honeybee foraging in the delicate, almost transcendent, beauty of the cherry blossom. The symbiotic relationship of the bee and the blossom inspires our divine spirituality to ponder how all living things are part of a magnificent web of interdependence and to appreciate how the natural laws that govern the ebb and flow of the cosmos reflect profound wisdom on the part of its creator.

It is our aesthetic spirituality that inspires us to create beauty through artistic endeavour and motivates us to adorn our bodies with jewellery and other items that enhance our personal attractiveness. We add aesthetic adornment to everyday utensils, tools, and machines, and even our weapons. Aesthetic spirituality also motivates us to collect beautiful objects in nature and to keep them nearby, so we can enjoy their beauty whenever we like. We capture and preserve the fleeting beauty observed in nature through our paintings, carvings, and sculptures, and we stir our emotions with music, dance, poetry, and other forms of artistic expression.

The accumulation of scientific knowledge and technology appears to have begun long before spiritual development, as evidenced by the two-million-year-old stone tools of *Homo habilis* and the invention of controlled fire by *Homo ergaster* about one million years ago. For the first one hundred thousand years or so after their arrival, *Homo sapiens sapiens'* cultural activities were, for the most part, similarly utilitarian, focused primarily on improving weapons for hunting ever-larger animals and developing more effective tools for butchering them. During this time, survival was a full-time occupation, leaving little time for the more aesthetic cultural activities of artistic expression or the pursuit of spiritual enlightenment.

The development of human culture was, and still is, a cumulative process, starting slowly and gradually gaining momentum until it suddenly accelerates rapidly. In mathematical terminology, such cumulative growth is referred to as *exponential*, and it seems that virtually all aspects of human cultural endeavour, including artistic expression, knowledge accumulation, technological innovation, and even spirituality, developed and continue to develop exponentially.

The pace at which our culture arose was closely linked with population growth, or more specifically, population density, since it advanced more quickly in those specific areas of the world where larger numbers of people lived together for lengthy periods. Knowledge, whether scientific or spiritual, is accumulated as new ideas and theories spring from those previously acquired. The higher the population density, the larger the knowledge base and the greater the amount of human ingenuity that is applied to exploiting it. Hence, the rate at which new technology and cultural elements are invented and refined is greater in densely populated areas.

Like cultural advancement, population growth displays a roughly exponential behaviour, although deviations occur in particularly harsh environmental conditions or after catastrophic geological events when human population decreases for some time. Such events are known as evolutionary bottlenecks. In the case of *Homo sapiens sapiens*, bottleneck events occurred approximately 170 thousand years ago, when environmental conditions on the African savanna were particularly harsh, and approximately 73 thousand years ago, following the enormous volcanic eruption of Mount Toba in Indonesia.

Human cultural development and population density were both minimal for the first 100 to 130 thousand years of Modern Human existence, but starting about one hundred thousand years ago, both began to accelerate. And between seventy and thirty thousand years ago, there was an explosive surge in both cultural sophistication and certain localized population densities, and both continued to grow exponentially through the advent of agriculture, continuing largely unabated into modern times. And it was during this cultural surge, according to most paleoanthropologists, that *Homo sapiens sapiens* transitioned from an archaic sapien into an anatomically and behaviourally, fully Modern Human.

So rapid was the cultural transformation between seventy and thirty thousand years ago that some paleoanthropologists refer to it as a *cognitive explosion*, and a few even attribute it to a favourable genetic mutation that occurred in human populations just prior to this time. More likely, however, the cultural blossoming was simply due to the exponential growth in population density that occurred at a few select sites throughout Africa and Eurasia. The increasing population density fuelled a similarly exponential growth in cultural and, in particular, artistic endeavours.

The Origin of Aesthetic Spirituality

"The calling of art is to extract us from our daily reality, to bring us to a hidden truth that's difficult to access—to a level that's not material but spiritual."

...Abbas Kiarostami

It cannot be known for certain when artistic expression first appeared, but there are some indications that it may have been as early as three hundred thousand years ago, prior even to the appearance of *Homo sapiens sapiens*. And the evidence also indicates that the first form of artistic expression was personal adornment.

Early humans, it seems, prized ochre, a naturally occurring mixture of ferric oxide and varying amounts of clay and sand that ranges in colour from yellow to deep orange, red, or brown. Pieces of ochre or ochre powder have been found in many ancient cave sites, often located several kilometres away from the nearest source. Since ochre was too soft to have had any utilitarian use in Palaeolithic times, most archaeologists agree that it was used to make decorative paint by grinding pieces of ochre into a fine powder and mixing it with water or rendered animal fat.

Cave sites throughout Africa and Eurasia provide much evidence that ochre-pigmented paint was used to adorn the bodies of the deceased as part of burial ceremonies and rituals, possibly as early as three hundred thousand years ago. Archaeologists speculate that it was also used by the living for personal adornment, beginning about the same time.

The earliest concrete evidence of ancient humans using ochre dates to about 285 thousand years ago at a *Homo erectus heidelbergensis* site in the Kapthurin formation of Kenya, where archaeologists have found about five kilograms of ochre pieces that had been collected and carried to caves located some distance from the source.

Evidence of ochre use has also been found at multiple Neanderthal sites throughout Eurasia, the oldest being the 250-thousand-year-old concentrates found at Maastricht-Belvédère in the Netherlands. And at Blombos Cave in South Africa, archaeologists unearthed evidence of Homo sapiens sapiens' use of ochre dating to one hundred thousand years ago.

Hence, it would appear that body painting and the beginnings of aesthetic culture and spirituality predated the arrival of *Homo sapiens sapiens*. In fact, it appears the practice began with late-tenure *Homo erectus* and was likely passed down to all of the sapiens subspecies, including the Neanderthals, Denisovans, and the various members of the ATG.

Starting about one hundred thousand years ago, personal adornment began to take on a new dimension when our ancient ancestors started to fashion bead necklaces, pendants, and bracelets.

Remnant strings of seashell beads have been unearthed at such widely scattered cave sites as the Skhul Cave in Israel, dated to approximately ninety thousand years ago; a site near the village of Taforalt in Morocco, dated to eighty thousand years ago; and the Blombos Cave in South Africa, dated to approximately seventy-five thousand years ago. At almost all sites, the beads were fashioned from the same genus of Nassarius marine snail shells and were all perforated for stringing in much the same manner.

While Nassarius shells are sometimes found pierced by natural predators, most of the beads found at archaeological sites appear to have been deliberately perforated in the same part of the shell, and some show signs of wear inside the perforation, indicating that they were likely strung together. It is believed that strings of the shells were worn as jewellery or possibly sewn into hide clothing as decoration.

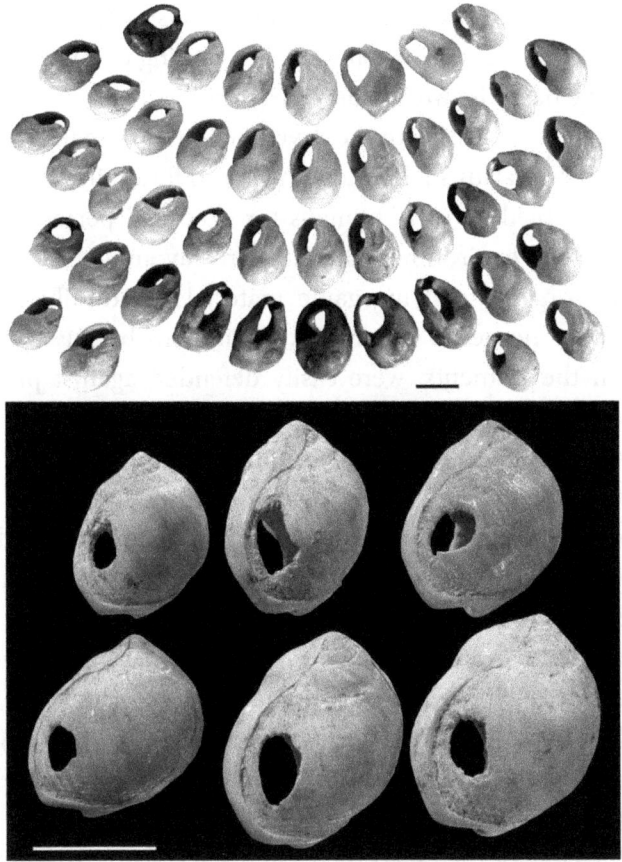

Nassarius shell beads found at Blombos Cave.

Image courtesy of Professor Christopher Henshilwood.

It is intriguing that the same shell beads were in such widespread use for over such a long period. Collections of Nassarius shells have been found hundreds of kilometres from any possible source, indicating that they were kept as personal possessions and carried, or worn, over long distances. Many other types of shells, as well as animal teeth or bone fragments, would also have made fine jewellery, but Nassarius shell beads appear to have been a favourite for thousands of years. While no evidence has been found to explain their popularity, Nassarius shell necklaces may have been a form of social statement or a long enduring fashion amongst early human groups. Since there is also evidence of trade, dating to over one hundred thousand years ago, amongst bands of *Homo sapiens*, it may be that Nassarius beads were collected as a trading commodity. If so, they may even have been the world's first currency.

Most artifacts pertaining to the development of human culture, including ochre and Nassarius shell beads, have been found in caves. Caves were frequently occupied by early humans for extended periods of time, and caves also created environmental conditions favourable for the preservation of both organic and inorganic materials. Naturally formed caves were a favoured homesite for prehistoric humans because they provided shelter from the elements, were easily defended against predators, and were a secure and comfortable place to sleep, prepare and cook food, socialize, store supplies, and craft weapons and tools. Habitable caves that were located close to fresh water and abundant food resources were much sought after and strenuously defended. Early humans occupied such caves for many generations, abandoning them only when the surrounding area became depleted of the life-sustaining resources they required.

While aesthetic culture appears to have taken root at several locales throughout Africa, the Middle East, and Europe, evidence suggests that it was in the caves of South Africa that human culture first began to truly take root, and *Homo sapiens sapiens* began the transition to behaviourally Modern Humans.

Along the edge of the escarpments and sea coast in southern Africa, a few scattered caves show evidence of early hominin occupation for tens, even hundreds of thousands of years, having human remains and artifacts preserved in the stratified sediment that composes the cave floors. Over millennia, as successive generations came and went, a chronological record of human cultural evolution was laid down, layer upon layer. And in the layers dated roughly seventy thousand years ago, copious amounts of newly invented artifacts have been found, reflecting a surge in human cultural development that began about that time. One cave site that exhibits these traits is located at Pinnacle Point, a small promontory immediately south of Mossel Bay, a town on the southeast coast of South Africa. Recent excavations of a series of caves at Pinnacle Point have found evidence of human occupation starting 170 thousand years ago.

The Pinnacle Point caves contained a large quantity of small blade tools and points, or *bladelets*, which would have been ideally suited for making smaller hafted weapons, dated to approximately seventy-one thousand years ago. Some interpret the finds to suggest that the occupants of the caves may have used slings, atlatls, or even bows as projectile devices for hunting smaller game in the more densely vegetated, inland areas. Some evidence suggests that the bladelets had been heat-treated with fire as part of the production process to improve the quality of the end product.

Not far to the west of Pinnacle Point, along the South Africa coastline, is Blombos Cave, an archaeological site that contains a large quantity of artifacts left by early human bands dating from about one hundred to seventy thousand years ago. These artifacts reveal much about the early phase of humankind's cultural and behavioural evolution.

Like the Pinnacle Point caves, archaeological material and faunal remains recovered from Blombos Cave show how early humans adapted to a new ecosystem by developing new technology, diversifying their diet, and modifying their hunting and foraging strategies.

The faunal record from Blombos Cave shows that occupants harvested coastal as well as terrestrial resources, and they expanded their diet to include fish, shellfish (giant periwinkle, limpets, and mussels), birds, tortoise and ostrich eggs, and mammals of various sizes. They hunted eland; gathered, collected, or trapped tortoises, hyraxes, and rats; and scavenged seal, dolphin, and whale meat from the beach. Judging by the amount and type of fish bones that were found in the cave, it is probable that the occupants used baited, bone hooks; barbed spears; or tidal traps to harvest fish.

The occupants of Blombos Cave also developed a more refined and effective spear and blade point technology known as the Still Bay tradition, named after the site of Stilbaai, located along the coast not far from Blombos Cave, where the artifacts were first found. The Still Bay tradition is broadly similar to the Mousterian tradition developed by the Neanderthals in Europe and by *sapien* groups elsewhere in Africa, but it shows an increased level of refinement and craftsmanship.

Still Bay points are formed by knapping flakes from both sides of the stone, which yields a slimmer, elliptic or willow-leaf-shaped point with sharp edges all round and pointed at both ends. Crafting of stone points in the Still Bay tradition requires much skill, and most of the artifacts found in the cave were production failures. The craftsmen at Blombos Cave preferred silcrete and quartz as raw material, and there is evidence that the stones may have been heat-treated during the production process to improve the ease of flaking and the quality of the end product. After the production process, the Still Bay points were hafted to wooden shafts and handles to make hunting spears, axes, and knives. They were so finely crafted and aesthetically pleasing that some archaeologists have suggested that their crafter must have taken considerable pride in their workmanship.

Still Bay bifacial stone points found at Blombos Cave.

In addition to the stone points, several bone tools, including fine awls and polished points, have been recovered from Blombos Cave. The awls were primarily made of fragments of long bone and would have been useful for piercing soft material such as leather or wood. As with the stone points, the bone points were carefully shaped and polished in an apparent attempt to improve their aesthetic appeal. Some bone tools were engraved with eight parallel incisions, which would have only been decorative as they would provide no additional function.

What appears to have been an ochre-processing workshop was also found at Blombos Cave. The two toolkits contained ochre, bone, charcoal, grindstones, and hammer-stones along with two abalone shells with which liquefied pigment mixtures were produced and stored. The toolkits represent the first known instance of deliberate planning, production, and curation of a pigmented compound, and its storage in a container suggests conceptual and cognitive abilities that are previously unknown for this time and serves as a significant milestone in the early evolution of the technological and cognitive abilities of Homo sapiens sapiens. More than eight thousand pieces of ochre material have been recovered from Blombos Cave, and some were incised with the cross-hatched design in combination with parallel lines, representing rudimentary abstract artwork. Cross-hatched engravings similar to those found at Blombos Cave have been found at other cave sites throughout southern Africa, indicating that there was a spatial and temporal continuity in the production and use of conventional symbols in the region.

Engraved ochre from Blombos Cave.

More than seventy Nassarius shell beads have been found at Blombos Cave, and studies show that the shells were more skillfully crafted than those found at other sites. The consistency in shell size and the colour of the bead clusters indicate that they were carefully selected and pierced with a fine point as the perforations are small and regular in shape. Wear patterns on the shells are evidence of stringing and consistent with use as personal adornment in the form of necklaces or bracelets. Some of the shells also show evidence of tinting with ochre pigment.

It appears that Blombos Cave was vacated rather suddenly about seventy thousand years ago as no artifacts dating to a later time have been found. The abandonment of what appears to have been a long-established, almost ideal dwelling place suggests that there was a major climate-change event that altered the region's flora and fauna to such an extent as to render human life unsustainable.

The Toba catastrophe, an enormous volcanic eruption that occurred about seventy-four thousand years ago at the site of present-day Lake Toba, on the island of Sumatra, Indonesia, is one possible such event.

The Toba super-eruption is the largest volcanic eruption ever known to have occurred on earth. It was roughly one hundred times greater than the 1815 eruption of Mount Tambora, the largest volcanic eruption in recent history, also in Indonesia. It is estimated that Toba's eruption deposited an ash layer about fifteen centimetres thick over much of south Asia and spewed a blanket of volcanic ash over the Indian Ocean as far west as the east coast of Africa. The volcanic ash deposits also extended northeast along the coast of present-day Vietnam to the South China Sea.

The ash and toxic gases reached high into the upper atmosphere and spread across the globe, causing a global volcanic winter that is estimated to have lasted six to ten years. The eruption occurred at the onset of the last Pleistocene glacial period and would have suddenly and substantially altered the global climate, likely accelerating the onset, and increasing the severity, of the last Great Ice Age.

The Toba catastrophe had a major impact on the flora and fauna of south Asia and likely of southern Africa as well. Population sizes of many animal species, including humans, were much reduced in some areas, and many animal species—that were able—vacated the most heavily impacted regions in search of more hospitable habitats. The Toba super-eruption may have been the event that caused the demise or exodus of much of the flora and fauna in the region of Blombos Cave and forced the human population to migrate north towards East Equatorial Africa and possibly beyond.

Elsewhere in South Africa, however, human culture continued to blossom after the Toba explosion. By fifty thousand years ago, artistic expression and particularly personal adornment, began to take on a new level of sophistication as early humans began to fashion jewellery from materials including ivory, bone, and likely wood. And by about forty-five thousand years ago, a new favourite bling appears in the archaeological record in the form of finely crafted ostrich eggshell beads.

Ostrich eggs would have been treasured by early Modern Humans. They are still harvested by today's San Bushmen who use them for a variety of things, much like their ancient ancestors would no doubt have. A single ostrich egg provides a nutritious meal for an entire family; the large, strong shells make ideal vessels for storing water; and broken fragments can be fashioned into fine beads for making jewellery. In modern times, the San women craft fine, exotic ostrich shell jewellery for personal adornment and for bartering in the local markets.

Ostrich eggshell beads have been found at several archaeological sites, including South Africa's Border Cave, a site that has proven to be a treasure trove of human cultural artifacts, dating roughly from seventy to thirty thousand years ago, precisely the time period during which human cultural development was blossoming.

Border Cave is located in South Africa, just below an escarpment rim near the border with Swaziland. Close to seventy thousand artifacts, both stone and organic, have been recovered from Border Cave, including the remains of six humans and more than forty-three species of mammals, some of which are now extinct. Analysis of the remains and artifacts reveals an active stone-working industry that used a broad range of materials, including chert, rhyolite, quartz, and chalcedony as well as bone and wood. The abundance of ostrich egg and marine shell beads are evidence of a relatively advanced artistic culture, and the diversity of animal remains shows that the cave's early inhabitants had a protein-rich diet of bush pig, warthog, zebra, buffalo, and several species of smaller animals.

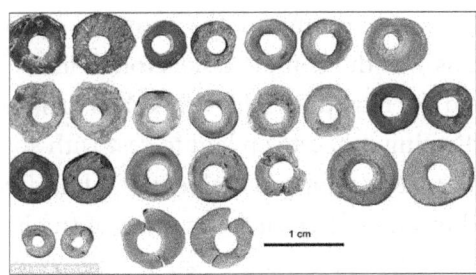

Ostrich eggshell beads from Border Cave in South Africa.

Picture courtesy of Lucinda Backwell, University of the Witwatersrand, Johannesburg, South Africa

Border Cave has a remarkably continuous stratigraphic record of occupation, but the artifacts found in the layers dated forty-five to forty thousand years ago indicate a rather sudden surge in human cultural development. Many of the artifacts dated to this period are still reflected in the culture of today's San Bushmen, leading archaeologists to conclude that today's Kalahari San people and the residents of Border Cave share a common heritage.

Artifacts found in Border Cave include bone points and awls, notched bones, wooden digging sticks, and quantities of both ostrich egg and marine shell beads. The bone points are of particular interest because they show signs of decoration in the form of spiral grooves filled with red ochre. Today, the San hunter-gatherers of the Kalahari personalize their arrowheads in much the same manner to identify ownership. The notched sticks found in the cave are also similar to San poison applicators, and some have been found with residue of a heated toxic compound still present, suggesting the Border Cave people were using poisoned arrow or spear heads by forty-four thousand years ago. The digging sticks were weighted with perforated stones in a manner also similar to San practice today.

Also recovered from the Border Cave is a small piece of notched baboon fibula, the oldest known artifact that shows evidence of notational tallying. Dated to thirty-five thousand years ago, the bone is incised with twenty-nine notches, similar to the calendar sticks used by the Namibian San today.

The Border Cave site produced the remains of five adult humans. Among the remains was the complete skeleton of an infant that had been buried in a grave in the fetal position, with a shell ornament and red ochre stains nearby, indicating it had been laid to rest with some ceremony. The remains have been dated to approximately one hundred thousand years ago. Ceremonial burial, even as simple as that of the Border Cave infant, indicates that the Border Cave people had at least an elementary humanity, given the dignity and respect with which they treated their departed loved ones.

Another important archaeological site in terms of Modern Human culture is Sibudu Cave, a large rock shelter located roughly forty kilometres north of the city of Durban, South Africa. The shelter is in a steep, forested cliff that overlooks the Tongati River, which now lies ten metres below the shelter. The shelter was occupied on and off by early bands of Homo sapiens sapiens from approximately seventy-seven to thirty-eight thousand years ago.

The artifacts and other finds at Sibudu Cave indicate that the people that lived there were highly advanced for their time. The finely crafted bone points and sewing needles are the earliest such tools found to date, and some scientists believe the points may have been used as arrow tips, in which case, it would be the earliest known evidence of the use of the bow and arrow.

It also appears that the residents of Sibudu Cave hafted stone and bone points onto shafts using glue, as opposed to sinew or twine, as early as seventy-one thousand years ago. Analysis of the points has revealed that they would have been attached to the ends of shafts using compound adhesives made of ochre and plant gum or, alternatively, fat mixed with plant material. Points were found with a cutting edge along their entire length, which means that they would have been attached to their hafts without twine. If that is so, the attachment would have called for a particularly strong adhesive glue.

To fix a flint or bone point to a wooden shaft using only glue is difficult, and today, only the strongest of synthetic glues would be suitable for such an application. To gain insight into the cognitive abilities of these early humans, scientists have replicated shafted tool manufacture using only methods and materials available at Sibudu. They have found that to make an adhesive strong enough to hold a point on a wooden shaft requires a very precise combination of substances, with exact ingredient proportions that then have to be heated to just the right temperature while avoiding boiling or dehydrating the mixture, either of which would weaken the mastic. Researchers conducting the experiment concluded that the people at Sibudu would have required the multilevel mental processes and abstract thought capabilities of a modern human to accomplish such a task.

Another intriguing find at the Sibudu shelter is evidence of the earliest use of plant bedding, dating back to approximately seventy-seven thousand years ago. The bedding consisted of sedge and various grasses topped with leaves from Cryptocarya evergreen trees and shrubs, which, when crushed, are pleasantly aromatic and contain traces of chemicals that have insecticidal and larvicidal properties, providing a measure of protection from mosquitos and other stinging insects. Various Cryptocarya species are still used as traditional medicines today.

A somewhat puzzling aspect of the artifacts found at the South African caves, in general, is the absence of evidence of two of the hallmarks of human culture: figurative art and music. To date, neither figurative carvings nor figurative wall paintings have been found in the South African caves dating prior to about ten thousand years ago, with the possible exception of the San Bushmen rock paintings. Similarly, no musical instrument artifacts have been found. One cannot discount the possibility that the artisans that lived in South African caves forty to thirty thousand years ago produced carvings and musical instruments from wood or some other perishable material, but there is no evidence to support such a hypothesis. For whatever reason, the first bone, ivory, and stone carvings; the first figurative cave art; and the first musical instruments in the form of bone flutes all first appeared in Eurasia between approximately forty and thirty-five thousand years ago.

The traditional music of Africa is ancient and diverse, with many distinct traditions. It is rich, vibrant, and often hauntingly beautiful. Songs and music are an important and integral part of African tribal culture, being used in rituals and religious ceremonies, to pass down stories from generation to generation, and for accompanying song and dance. Traditional African music today is performed on a wide variety of percussion instruments, including many types of drums like the djembe, the marimba, the xylophone, and the mbira but not on stringed or wind instruments. Indeed, no evidence of the fabrication of wind or stringed musical instruments has ever been found at an African archaeological site older than ten thousand years. Given the lack of such instruments in traditional African music, it would appear that wind and string musical instruments were not part of the prehistoric African musical culture and were only invented after prehistoric humans left the African continent.

WAKAN TANKA

The Origin of Divine Spirituality

"Our scientific power has outrun our spiritual power. (And so) we have guided missiles and misguided men."

...Martin Luther King, Jr.

The beginnings of cultural development reflect a dawning not only of aesthetic spirituality but of the divine aspect as well.

The fundamental element of our divine spirituality is the belief in a universal spiritual essence that resides within each of us, constituting the individual soul. The universal essence is associated with the creator of the universe, and it is our soul that gives us vitality, consciousness, and the awareness of self.

One can only speculate on how the awareness of our divine spirituality began. Perhaps an ancient ancestor began to ponder the natural world's division into animate and inanimate entities and wondered about the mysterious transition that overcame a living organism as it succumbed to disease or injury and died. Where one minute there was vitality, a certain animating energy, in the next instant it was gone, and a once vibrant animal was suddenly still and lifeless. Such a realization would have been most poignant when a life partner, child, or close friend died. But all animals displayed the same vitality transitions at birth and at death, and even the plants that were harvested for food were strong and vital when plucked from the ground but became limp and withered within a short while.

This mysterious vitality imbued every plant and animal with a vigorous *life force,* yet had no form in itself. It couldn't be seen, smelled, heard, or felt, and the only evidence of its existence was the marked difference in a plant or animal between when it was present and when it was absent. Just as many people do today, our ancient ancestors would have wondered about the nature of this mysterious vitality—where it came from at birth and where it went after death.

Some thoughtful ancestor would also have noticed that all living things were animated by the same vitality, which embraced them in an intricate web of interdependence, of life and death. All living things depended upon one another for survival. Grazing animals relied on living plants for food, without which they perished and their decayed remains became nourishment for the very plants they had depended upon for food. The great cats and other predators that roamed the savanna were nourished by the flesh of ungulates, and human life depended on a variety of plants and animals for food, clothing, tools, and shelter.

As early humans pondered these great mysteries, they began to develop an understanding of the interdependence of all life and how the natural world functioned in an ordered and highly integrated fashion. Eventually, they came to appreciate the profound wisdom it reflected, and this appreciation led to a reverence for this all-pervading, immanent vitality. They began to associate it with the creator of the world and to refer to it as the *Great Spirit*.

Gradually, the belief in the Great Spirit evolved into animism, the belief that the Great Spirit also assumed an individual identity in all living things, referred to as the *inner spirit*. In more modern times, this inner spirit became known as the divine within or the *soul*. People came to speak of their own inner spirit and the spirits of the animals with which they shared their environment—the spirit of the cave bear, the eland, or the woolly mammoth. The concept led to the belief in the existence of a spiritual realm where the Great Spirit and the spirits of all living things resided.

From such beginnings, human spirituality blossomed and, over time, evolved into the many spiritual and religious belief systems found throughout the world today. But the ancient concepts of a universal essence or spirit and the individual soul have persisted over the millennia. Virtually all modern religions are premised on a belief in the existence of the individual soul and an omnipresent supreme being or god, which most associate with the creator of the cosmos.

Many paleoanthropologists associate ceremonial burial and other signs of compassionate behaviour with the presence of a nascent divine spirituality in prehistoric human societies. Some interpret such evidence as indicating a belief in an existence after death, while others suggest that, at best, it indicates emotional attachment and respect for the other members of a social group. At a minimum, caring for the well-being of others shows a developing sense of humanity, and the burial of a departed friend or family member with dignity and respect implies at least an intuitive awareness of a departed soul. And both actions are indicative of an emerging divine spirituality.

Deep inside a cave system in the Atapuerca Mountains of northern Spain, archaeologists have discovered a pit that contains the remains of at least twenty-eight ancient hominins. The site, known as Sima de los Huesos, or the *pit of bones*, is suspected by some to have been a burial site where the deceased were laid to rest with some ceremony. The pit also contained the deformed skull of a twelve-year-old girl that would have been severely disabled from birth, suggesting that others in her social group had cared for her during her short life.

Remarkably, researchers were able to extract enough DNA material from a thigh bone found in the pit to produce a near-complete mitochondrial genome. The DNA analysis showed the individual lived about four hundred thousand years ago, making it the oldest hominin DNA sequenced to date.

Perhaps even more startling, a comparison of the DNA with that of Neanderthals and Denisovans indicated similarities with both, and it likely belonged to an ancient common ancestor. The conclusion is that, like our aesthetic spirituality, our humanity and divine spirituality may have had their beginnings with late-tenure *Homo erectus*, even before the appearance of any of the sapien subspecies. And from *Homo erectus*, it appears to have been passed on to the sapiens as there is much evidence that both the Neanderthals and early *Homo sapiens sapiens* both practiced ceremonial burial.

At the Kaprina cave site in Croatia, the remains of several dozen Neanderthal individuals, ranging in age from two to forty years, have been unearthed, which makes Kaprina the largest and richest discovery of Neanderthal remains ever found. The site has been dated to approximately 130 thousand years ago, and the archaeological evidence indicates that the cave contained both living quarters and a community burial site. The Kaprina evidence, along with evidence of ceremonial burial at other Neanderthal sites such as Shanidar Cave in Iraq and La Chapelle-aux-Saints in southwestern France, leads archaeologists to the conclusion that the Neanderthals routinely practiced ceremonial burial throughout their range in Europe and western Asia.

Amongst the earliest evidence of ceremonial burial by *Homo sapiens sapiens* has been found at the Qafzeh and Es Skhul Caves in Israel, dated to approximately one hundred thousand years ago. One such burial is of a ten-year-old-boy who was buried in a rectangular grave that was carved out of the bedrock. The skeleton was laid on its back, with the legs bent to the side and both hands placed on either side of the neck. The antlers of a large, red deer had been placed in its hands, which clasped them to the chest. In addition, seventy-one pieces of ochre were found, some of which were associated with the burial of a brain-damaged child. Red, black, and yellow ochre-painted seashells were also found at the site.

The earliest evidence of a belief in a spiritual realm was the discovery of anthropomorphic art that depicts figures of a human body with an animal head. Anthropologists associate anthropomorphic paintings and figures in prehistory as signs of shamanism, a spiritual ritual practiced by a tribal healer or shaman. Typically, the shaman would enter an altered state of consciousness, enabling his own spirit to enter the spirit world, commune with animal spirits, and channel transcendental energies into this world for divination and healing emotional and psychological disorders. Anthropomorphic figures represent the shaman's spirit combined with that of an animal, typically an animal that had a special spiritual relationship with the shaman and the society in which he lived.

The oldest anthropomorphic figurine, known as the Löwenmensch, or Lion Man, has been dated to between forty and thirty-five thousand years ago. It was found at the Hohlenstein-Stadel cave site in the Lone River valley of southern Germany. The figure was carved from mammoth ivory, stands about thirty-one centimetres tall, and depicts a human body with the head of a lion. The fragments of similar figures have been found at neighbouring sites, suggesting that the cave lion, which lived throughout Eurasia at the time, held some spiritual significance to the shaman and the people in the region.

The Löwenmensch.

Our divine spirituality is closely tied to our sense of humanity, and it is likely that the two evolved in concert, stemming from our emotional fabric that evolved over millions of years. Just as our socially bonding emotions appear to have their origins with the Australopithecines, so too may the faint beginnings of human spirituality been manifest in these ancient hominins.

Our divine spirituality also influences many of our artistic endeavours. In modern times, much of our art, sculpture, music, and literature is inspired by our religious and spiritual beliefs. One need only visit the Louvre museum in Paris or the Sistine Chapel at the Vatican in Rome to see how religious belief has inspired some of the most beautiful paintings and sculpture ever created.

It is also our divine spirituality that guides us to use our scientific knowledge and creativity for the betterment of humankind. Technology in itself is morally and ethically neutral, and many technological inventions have a multiplicity of uses, some of which are constructive and beneficial, and others of which are destructive and harmful. A hafted spear can be used to kill animals for food, but it can also be used to slay a fellow human being. Enriched uranium can be used as fuel in a nuclear electricity generation plant, but it is also the essential component of a nuclear bomb.

Sadly, since the dawn of civilization, humankind has applied much, if not most, of their advanced technology to pursue wealth and power at the expense of others. The conquering, killing, and subjugation of our fellow human beings has been the main purpose of most of our technological development in modern times. The technology used in most of the machines and devices we all use and enjoy today—computers, smartphones, the Internet, automobiles, and airplanes—was originally developed for military use. Today, about 60 percent of the discretionary component of the United States' federal budget is spent on its military, and many countries spend even more on a percentage basis.

Such a sad reality has prompted some, like Martin Luther King, Jr., to argue that our scientific knowledge is outpacing our spiritual development, and we are spirituality ill-equipped to responsibly control or use the advanced technology that we can produce. In modern times, our spirituality is being suppressed while our scientific knowledge and technology continue to advance at a bewildering rate. We have reached the point where our weapons can destroy not just our perceived enemies but most of the flora and fauna on the entire planet, including ourselves, many times over. Yet we still live on the brink of using such weapons to impose our will on each other.

How can it be that the same marvellous mind of *Homo sapiens sapiens* could compose music that resonates with our souls so profoundly that our hearts are lifted to untold heights of rapture, and still devise technological devices that can murder hundreds of thousands of fellow human beings with the mere flick of a switch?

Such is the enigma that is the human mind.

CHAPTER EIGHT
Leaving Africa

"If you want to travel fast, travel alone, but if you want to travel far, travel together."

...African Proverb

FOR THE FIRST one hundred or so thousand years following their appearance on earth, the various sapiens subspecies throughout Africa and Eurasia lived largely in isolation, unaware of one another's existence. Indeed, two hundred thousand years ago, it would have appeared that the stage was set for several sapiens subspecies to continue to evolve independently into modern times, with the Neanderthals in Europe, the Denisovans in Asia, and several subspecies widely dispersed throughout Africa. One can only imagine the cultural and social diversity that would exist throughout the world today if this had been how our modern societies evolved.

But it was not to be.

In contrast to the wide genetic diversity of African tribal people, those of us living outside the African continent today have surprisingly little DNA diversity, and genetic analysis has shown that all people throughout the globe, except for African tribal people, are descended from a relatively small group of Modern Humans that left Africa, starting about seventy thousand years ago. From Africa, they spread across Eurasia and eventually into the Americas. By fifteen thousand years ago, bands of Modern Humans could be found on every continent of the world save Antarctica.

The continent of Africa is quite isolated from the rest of the earth's land masses, and as early humans began to leave the continent, there were only a few routes open to them. During the Pleistocene, there were only a few interglacial periods when the climate was conducive to migration and settlement in the more northern regions.

The only land route out of Africa, both today and during the Pleistocene epoch, is the Levantine corridor, a narrow strip of land between the Mediterranean Sea to the northwest, the Red Sea to the southwest, and deserts to the southeast. This corridor has provided a land route for migrations of animals between Eurasia and Africa for millions of years, and it was no doubt used by *Homo erectus* bands when they wandered out of Africa some two million years ago.

Bab el-Mandeb Strait, which lies between Djibouti and Yemen, is about thirty kilometres wide, and there are islands in the strait dividing the passage into two channels of three kilometres and sixteen kilometres. Evidence shows that early humans crossed the strait from Africa onto the Arabian Peninsula when sea levels were lower, making the crossing even shorter than it is today.

During the Pleistocene glacials, the Sahara and Sinai regions were most frequently cold and arid, and most animals, including early humans, would have had no reason to cross them into the Middle East and beyond. During these times, the area of the Levant and Arabian Peninsula was populated largely by Arctic flora and fauna that had migrated southward ahead of advancing glaciation in the north and amongst them, the Neanderthals.

But, during the interglacials, the climate in the region was warmer and wetter, rivers and lakes filled, the deserts greened, and forests expanded. The Arctic fauna moved northward again, to be replaced by tropical fauna, including the *Homo sapiens sapiens* that crossed the Levantine corridor from Africa. During such times, the region of the Levant essentially became an extension of North Africa.

During the last two hundred thousand years, there have been two major ice ages, with the first lasting from about 190 to 130 thousand years ago and the second from 115 to 12 thousand years ago. During these glacial periods, much of Europe was covered with ice and snow, and the climate in the southern regions, including North Africa and the Middle East, was cold and dry. During the most extreme periods, the earth's average temperature dropped to levels 5 to 7°C lower than today. Sea levels dropped, forests disappeared, rivers dried up, and much of central Europe and Eurasia became steppe grassland. Deserts appeared in North Africa and the Middle East, and the east African savanna cooled and withered.

Between the two glacials, however, there was an interglacial where the earth became rather warm and wet, unusually so, at least for the Pleistocene epoch. The interglacial, known as the *Eemian period*, persisted for approximately fifteen thousand years, between 130 and 115 thousand years ago. During that time, the earth's average temperature spiked, reaching levels approximately 3°C warmer than today. The warming climate caused rainfall to increase, which created green, hospitable ecosystems in the Sahara, Levantine corridor, and Arabian Peninsula. Rivers flowed northward across the Sahara, creating lush wetlands that teemed with tropical flora and fauna. Tropical forests and wetlands extended across North Africa into the Levant and Arabian Peninsula while boreal forests expanded throughout Europe and Asia as far north as the Arctic Circle.

During the Eemian, African animals and bands of *Homo sapiens sapiens* crossed the Levantine corridor and fanned out into the Levant, the wetlands of the Tigris–Euphrates Basin, and south into the Arabian Peninsula. As these tropical fauna entered the region, the Neanderthals, along with other Arctic animals, migrated northward, abandoning the Levant and returning to central and northern Europe.

Evidence of Modern Human migration via the Levantine corridor during this time has been found at two archaeological sites in Israel: the Mugharet es-Skhul cave site on the western slopes of Mount Carmel, about three kilometres south of Haifa, and the Jebel Qafzeh cave near Mount Precipice, just south of Nazareth. Artifacts and remains found at both sites have been dated to the Eemian period, and at both sites, most of the human remains recovered indicate some sort of ceremonial burial. They appear to have been laid to rest with grave goods, which, in more modern times, are interpreted as items placed on or beside the deceased to smooth their journey into the afterlife or provide offerings to the gods. However, the extent to which the grave goods found at the Mugharet es-Skhul and Jebel Qafzeh caves were laid with such intent cannot be known with any degree of certainty.

In addition to the burials, a series of hearths, stone artifacts, and animal bones have been found, along with evidence of more advanced cultural development in the form of Nassarius shell beads tinted with red ochre. The technology of the people at both sites was Mousterian, with points and blades manufactured using the Levallois technique. The degree of cultural development exhibited by the people at the Levantine sites was roughly the same as that found at African sites dating to the same time period.

There is evidence that near the beginning of the Eemian warming, when sea levels were still low, early humans crossed the Mandeb Strait from the Horn of Africa, entering the Arabian Peninsula via Yemen and fanning out into Oman, Saudi Arabia, and Iran, eventually reaching as far east as south-central India. For example, in the Dhofar region of southern Oman, archaeological sites have been discovered that show that bands of *Homo sapiens sapiens* were living along the banks of rivers and lakes in the area from the beginning of the Eemian period. Further east, in a rock shelter just south of the Straits of Hormuz, at the base of Mount Jebel Faya in the United Arab Emirates, archaeologists have discovered early human artifacts dating from 127 to 95 thousand years ago.

Immediately following the Eemian period, approximately ninety thousand years ago, the earth entered another severely cold period known as a Heinrich event, which was followed by the onset of prolonged glaciation. The Sahara, Levant, and more northerly Sinai regions turned cold and dry once again, and the tropical flora and fauna, including the early humans, disappeared from the region. They were replaced, once again, by more northern flora and fauna, including the Neanderthals. Whether or not the early humans migrated back to Africa cannot be known. However, the total Levantine population during this time is estimated to be less than five hundred individuals. They would have been barely viable, leading most archaeologists to believe that they most likely perished about ninety thousand years ago.

Those people in the southern regions of the Arabian Peninsula, however, appear to have fared better than their more northerly neighbours, as evidenced by the discoveries at Jwalapuram, an archaeological site in the Jurreru River valley of Andhra Pradesh in southern India.

Artifacts recovered from Jwalapuram include blades and scrapers along with the cores from which they were crafted. A piece of red ochre that shows signs of scraping was also found. Unfortunately, no human remains have been found, so it cannot be known whether the people that lived there were early Modern Humans or some other sapien subspecies. But most researchers believe they were *Homo sapiens sapiens*.

The Jwalapuram site is intriguing because similar artifacts were discovered immediately above and below the fifteen centimetre layer of ash that was deposited over the area during the Toba eruption, seventy-four thousand years ago. This circumstance suggests that the humans living at the site may have survived the eruption. However, if they did survive the eruption, the people at Jwalapuram appear to have left the region shortly thereafter as no further trace of them has been found throughout India dating to the same general time frame.

Immediately after the Toba catastrophe, there is a hiatus in the archaeological record of the Middle East and India from seventy-four to approximately sixty-five thousand years ago. It seems quite likely that the fallout from the Toba eruption, in combination with the cold and dry conditions brought on by the developing Pleistocene glacial event, proved too much for the human populations living in southern Eurasia at the time. They most likely perished, along with much of the flora and fauna upon which they depended for food.

Following the Eemian interglacial period, the earth entered the last Pleistocene ice age that continued to intensify until reaching its apex between twenty-five and fifteen thousand years ago. However, the gradual onset of the ice age was not uniform. Rather, there were short periods when the climate was less severe, and temperatures reached levels of only 1 or 2°C lower than today. It was during one of these warmer periods, between seventy and sixty-five thousand years ago, that the next wave of *Homo sapiens sapiens* left the Africa continent. And all evidence points to the likelihood that the first migrants came from the region around the Horn of Africa, and their exodus was via the Bab el-Mandeb Strait.

It cannot be known for certain what prompted our ancient ancestors to leave the African continent. Perhaps a brief period of climate change rendered their environment uninhabitable, or conversely, perhaps the people living along the coast of the Red Sea and the Horn of Africa were flourishing, and their numbers expanded to the point where local resources became depleted.

Whatever the reason, sometime between seventy and sixty-five thousand years ago, a group or groups of *Homo sapiens sapiens* individuals crossed the Bab el-Mandeb Strait to begin an incredible journey of migration and settlement that would see their descendants ultimately reach all the major land masses of the earth over the next fifty thousand years. Scientists don't agree on how many there would have been—some estimate a few hundred, others say a few thousand—but the genetic evidence indicates that the number was relatively small.

Over subsequent generations, these early migrants continued to move eastward along the southern coast of Arabia, settling whenever they found hospitable places with good shelter and ready access to the sea's bounty. They appear to have thrived; their numbers grew, and within a few millennia, bands of *Homo Sapiens sapiens* could be found living all along the coast of the Arabian Sea.

Initially, as the people wandered ever eastward, they would have settled close to the coast. They were fully adapted to hunting and gathering marine food resources, and there was no need to change that. Inland, the climate was cold and dry as the ice age advanced, and land-based food resources were harder to come by, whereas each new coastal plain or river estuary they encountered would have offered a fresh new bounty of marine resources for the harvesting.

These early migrants must have built watercraft to better enable them to harvest the sea's bounty and to move along inhospitable sections of coastline. Over time, they developed ever-larger watercraft that could transport them in larger numbers, together with sufficient food, water, and other supplies that they needed to traverse longer stretches of coast or to travel to offshore islands that showed promise as agreeable places to live.

Eventually, during times when environmental conditions were favourable, some bands began to disperse inland, settling in valleys along major rivers like the Indus and Narmada or at river estuaries such as the Ganges-Brahmaputra Delta region. As bands dispersed into the interior of their newly adopted homelands, some readopted a land-based hunting and foraging culture, and over time, after living in isolation from one another, they developed divergent cultures with different languages and dialects, customs, and hunting practices. The rich diversity of indigenous people and culture found in southeast Asia today began to emerge.

These prehistoric wanderers were not on a mission of exploration; rather, they moved into new territories only when their population grew to the point where it became necessary to find new locales in which to live. Over the centuries, their wandering took them ever eastward—along the shores of the Arabian Sea, around the coastline of the Indian subcontinent, and into southeast Asia, reaching present-day Indonesia sometime between sixty and fifty thousand years ago.

During the last Pleistocene glacial, sea levels reached a minimum at approximately sixty-five thousand years ago. At that time, the region encompassing the Malay Peninsula and the Indonesian archipelago was connected by the exposed Sunda continental shelf, which composed part of a large peninsular region known as Sundaland. Similarly, the exposed Sahul continental shelf to the south formed a greater Australian land mass known as Sahulland, which included Australia, Tasmania, and New Guinea. Between the two was an archipelago collectively known as Wallacea, which included the islands of Lombok, Sumbawa, Komodo, Flores, Sumba, and Timor. These islands were separated by deep water channels and were never joined by land bridges to either Sundaland or Sahulland, even at the lowest Pleistocene sea levels.

The Philippines and the islands of Melanesia also remained separate during this time, although the distances between them were greatly reduced, and sea crossings from island to island were well within the capabilities of the first *Homo sapiens sapiens* that arrived in the region.

As the migrants entered the Sundaland Peninsula, they encountered bands of Denisovans that had migrated southward from the Siberian steppes, perhaps to escape the harsh winter conditions that prevailed in the north at that time. The encounter appears to have been at least somewhat amicable as an analysis of the DNA of people living throughout southeast

Asia and Australia indicates that Modern Humans and Denisovans interbred around this time frame. Between 3 and 5 percent of the DNA of today's Melanesians and the Aboriginal people of Australia is traceable to the Denisovans.

As the early human migrants entered Sundaland, they appear to have fanned out quickly, taking up residence at several locales along the coastlines as well as in the interior of the peninsula. Coastal sites have long since been destroyed by rising sea levels, but remains and artifacts found at the Niah Caves in the Malaysian State of Sarawak on the Island of Borneo have been dated to forty-six thousand years ago, and evidence found in the Tabon Caves of the Philippines indicates occupation by Modern Humans almost fifty thousand years ago.

The Negrito people of southeast Asia are the descendants of these first Sundaland immigrants. The term Negrito is used to describe several indigenous ethnic groups that share the same physical characteristics, most notably dark skin, tightly curled hair, and a diminutive stature. It includes the Andamanese people of the Andaman Islands, the Orang Asli of Malaysia, the Maniq people of Thailand, and the Aeta, Ati, and about thirty other ethnic groups living throughout the Philippines. Recent genetic studies of Negrito people have substantiated the archaeological evidence and confirmed that most have localized, but very ancient, DNA. The same genetic studies indicate that they began settling in the region as early as sixty thousand years ago, remaining more or less in the same locale since that time.

While most of the indigenous people of southeast Asia are adopting a modern lifestyle and are gradually being assimilated into modern societies, a few, such as the Orang Asli in the rain forest of peninsular Malaysia, still maintain a hunter-gatherer lifestyle similar to that practiced by their ancient ancestors. Today, about fifteen hundred individuals still inhabit the Taman Negara National Park in central Malaysia.

The Orang Asli live in familial groups with about ten families forming an encampment, composed of lean-to shelters. Each encampment generally has control of the land immediately around it, but since they do not believe in the concept of private land ownership, they consider themselves caretakers of the land, rather than its owners.

Traditionally, the Orang Asli subsisted primarily on gathered tubers, fruit, leaves, fish, and small game such as monkeys and civets, and they continue to do so today, although they now supplement their diet with modern staples such as sugar, tea, and rice. They are a nomadic people; once most of the usable wild food resources have been depleted from a given location, they move to another spot but always stay within their familiar habitat. They have their traditional encampment locations where they leave the pole frames of their shelters and a few other useful items as they move from one campsite to the next, living in one area while the food and resources are naturally restored in the others.

From Sundaland, the migrants began to make the sea crossing into the Islands of Wallacea, reaching the Island of Flores about fifty thousand years ago, where it appears they played a role in the extinction of both Flores Man and his favourite large prey, the Stegodon—a now-extinct species of elephant that lived throughout southeast Asia at the time. From Wallacea, they continued their sea voyaging, setting foot on the shores of Australia for the first time somewhere around fifty thousand years ago.

As these early Modern Humans began to wander along the coastlines and disperse into the interior of Australia, they encountered a world of fauna that was different than anything they had ever seen before. Much of the central and northern parts of the Australian outback consisted of semi-arid grassland and was home to a large number of megafauna. Like today, there were several species of kangaroo and wallabies, but they were generally larger, like the giant Procoptodon, a five-hundred-kilogram kangaroo that stood over two metres tall. There were wombat-like creatures such as Diprotodon that was about the size of a hippopotamus, and a 150-kilogram marsupial lion. Large reptiles included several species of crocodiles and a Volkswagen Beetle–sized terrestrial tortoise, Meiolania, that had a horned head and spiked tail. There were giant monitors like *Megalania prisca*, which grew to a length of almost eight metres.

There were also multiple species of emu-like flightless birds, including the largest, *Genyornis newtoni*, which stood over two metres tall. A favourite delicacy of early humans appears to have been *Genyornis newtoni*'s rather large, 1.5-kilogram eggs, which they cooked over an open fire. At more than two hundred sites across the Australian continent, pieces of burned or partially burned eggshells have been found, indicating that the new arrivals had developed a taste for cooked *Genyornis newtoni* eggs rather quickly and taken to raiding their nests whenever they could find them. Emu eggs were a delicacy as well, also preferred cooked.

It appears that it didn't take long for at least some of the new human immigrants to abandon coastal living and a diet of seafood in favour of a life on the open grasslands and a diet rich in meat protein. While there have been few archaeological sites found with quantities of slaughtered megafauna remains, most scholars agree that it is no coincidence that most—some estimate about eighty-five percent—of Australia's prehistoric megafauna, including *Genyornis newtoni*, were extinct by about forty-five thousand years ago, only a few millennia after the arrival of *Homo sapiens sapiens*.

One of the earliest archaeological sites on the Australian continent is located at Devil's Lair, a cave in the far southwest of the continent, about twenty kilometres north of Cape Leeuwin and five kilometres inland from the present coast. Remains found in the cave show that Modern Humans arrived there approximately forty-eight thousand years ago and were hunting the large land megafauna, including the giant kangaroos, Protemnodon and Sthenurus, that lived in the area at the time. Across the continent at Lake Mungo, the forty- to fifty-thousand-year-old remains of two individuals, dubbed Mungo Man and Mungo Woman, show evidence of ceremonial burial. Mungo woman appears to have been at least partially cremated, and the remainder of her body was crushed and meticulously prepared for burial, suggestive of ritual and some form of ceremony. Mungo Man's remains were covered in red ochre, a practice that is also believed to be part of a burial ritual.

In the rock shelters and caves of southwest Tasmania, artifacts and human remains have been found dating to as early as thirty thousand years ago. The Tasmanian sites are the most southerly archaeological finds dated to this period that show evidence of human occupation.

As population swelled throughout southeast Asia and Australia, the migrants began wandering northward from Sundaland, along the eastern coasts of Vietnam, Korea, and China. Artifacts found in cave sites in the region indicate that they reached the Korean peninsula as early as forty thousand years ago, northeastern Russia by thirty-eight thousand years ago, and Japan by thirty-five thousand years ago.

As in southeast Asia, some of these migrants would have settled in the lower reaches and estuaries of large rivers like the Yangtze and Huang He and, over time, migrated along these rivers to inland China. The Ma'anshan Cave site on the bank of the Tianmen River in the northwest Guizhou province of China contains artifacts and evidence of human occupation from thirty-five to eighteen thousand years ago. The cave artifacts are mostly bone-butchering tools, awls, and points fabricated by Modern Humans that came into the region either from the South China Sea coast or from southeast Asia via the Brahmaputra, Salween, and Mekong river systems that have their headwaters in the Tibetan Plateau.

It also appears that the early human migrants travelled inland along the Huang He and onto the Great China Plain between forty and thirty thousand years ago. Genetic evidence indicates that they reached the Zhoukoudian archaeological site, about 40 kilometres southwest of Beijing, around this time and interbred with the local, possibly Denisovan, indigenous population.

By about forty-five thousand years ago, Modern Human culture was beginning to blossom at many places throughout Eurasia and Africa as *Homo sapiens sapiens* transitioned to fully Modern Humans. It is interesting that societies of these early humans that were so widely separated geographically, for so many millennia, all began to exhibit similar cultural advances at more or less the same time.

The new immigrants in southeast Asia were no exception.

In a remote mountainous region of East Kalimantan, Indonesia, there is a network of limestone caves perched atop forbidding, densely forested peaks, and in these caves, there is a treasure of prehistoric artwork. Thousands of human hand stencils adorn the walls, made, it would appear, by blowing a mouthful of red ochre-based paint over an outstretched hand. In addition, the caves contain some of the earliest known examples of figurative art depicting cattle-like animals, most likely a species of Banteng. Both the hand stencils and animal paintings have been dated to about forty thousand years ago. The hand stencils associated with the animal paintings are of various sizes, so it appears that people of all ages enjoyed painting on the walls of caves forty thousand years ago.

Across the Macassar Strait, on the Indonesian island of Sulawesi, even more striking examples of figurative art have been found, dated to about thirty-five thousand years ago. The paintings adorn the walls of caves and rock shelters, located at the foot of spectacular limestone towers that rise up from the surrounding rice fields near Maros in the southwestern part of the island. The oldest are two paintings of animals, likely a babirusa, or pig-deer, and another animal, perhaps another species of wild pig or cattle. The animal drawings were done with red ochre and black charcoal-based pigments, and the artist appears to have used a fine bone or shaped wooden stick to apply them. Some of the Sulawesi animal drawings are exquisite, with remarkable detail and artistry, indicating that the artist was not only artistically cognitive but talented as well.

An example of cave art from the limestone caves of Sulawesi.

Archaeologists have long pondered the significance of prehistoric cave drawings. In many instances, caves containing elaborate works of art show no evidence of habitation, so it is believed they had spiritual significance and were used only for special occasions and ceremonies, reminiscent of today's religious cathedrals and shrines. Various species of wild boar, including the babirusa, were present throughout southeast Asia forty thousand years ago, and their meat was very likely relished by the people that settled there. Just as the San people of the Kalahari developed a mystical reverence for the eland, and the Lakota Sioux people of the American Western plains revered the buffalo, the indigenous people of Sulawesi may have come to associate themselves spiritually with the babirusa. The cave paintings may have been some sort of tribute to this association.

Alternatively, the paintings may have been produced by someone who just enjoyed painting and took pride in creating something beautiful for the enjoyment and appreciation of his fellow band members. We shall never know what motivated them to produce such intricate works of art, but it is clear that the individuals that made the Sulawesian cave art were as cognitively advanced as we humans today. Their art would be well received and appreciated by most art aficionados today.

CHAPTER NINE
The Mammoth Steppe

A HUNTER-GATHERER LIFESTYLE requires a large amount of land and natural resources to sustain it, and the larger the group, the more land is required. So as the population of southeast Asia continued to increase, the more hospitable areas would likely have become overpopulated, straining local supplies of resources. Eventually, the hunter-gatherer lifestyle would have become unsustainable for such large numbers of people.

Such population pressures would have provided the impetus for continued migrations, and the only available, unoccupied land lay inland from the coast and to the north. But to the north were the vast mountain ranges of the Alpine-Himalayan, or Alpide Belt, a massive series of orogenic mountain chains that extend along the southern margin of Eurasia, separating the southern and southeastern regions from central Eurasia. This was a formidable barrier, especially during times of extreme cold when glaciers extended into the valleys and icefields covered the high mountain passes.

North of the Alpide Belt, on the flanks of the Himalayas, lies the Tibetan Plateau. This plateau is a high-altitude, arid steppe that is interspersed with mountain ranges and large, brackish lakes. It stretches approximately 1,000 kilometres north to south and 2,500 kilometres east to west. With an average elevation exceeding 4,500 metres, the Tibetan Plateau is sometimes called the Roof of the World and is the world's highest and largest plateau, with an area of 2.5 million square kilometres.

The plateau is surrounded by massive mountain ranges and is covered with some ten thousand glaciers, which are the headwaters of several large rivers. Toward the eastern end of the plateau are the mountainous sources of the Brahmaputra, Salween, Mekong, Huang He, and Yangtze rivers, which flow for thousands of kilometres throughout southeast Asia before emptying into the Bay of Bengal, the Adaman Sea, the Gulf of Thailand, and the South China Sea. At the western end of the plateau are the headwaters of the Indus River, which flows from the vicinity of Lake Manasarovar, around the western end of the Himalayas, through Punjab and Pakistan before emptying into the Arabian Sea.

The estuaries and valleys in the lower stretches of these great rivers were occupied by human migrants as they moved along the coast. And as overpopulation pressures mounted, the rivers ultimately afforded them with a route further inland, around the Tibetan Plateau and into central Asia.

North of the Alpide Belt and the Tibetan Plateau lay a vast grassland plain, known as the Mammoth Steppe, so named because of the large elephant-like beasts that lived there but also because of its massive extent. Fifty thousand years ago, the Mammoth Steppe was the earth's most extensive land-based biome. It stretched from Spain eastward across Europe and Siberia into Alaska and the Yukon, where it ended against the massive icefields of the Wisconsin glaciation. North to south, it extended from the Arctic islands to central China where it came up against the icefields of the Tibetan Plateau.

At that time, much of the earth's fresh water was frozen in vast continental ice sheets like the Wisconsin glaciation and the Patagonian and Scandinavian icefields, which left little moisture in the earth's atmosphere. Precipitation, in the form of rain or snow, was low across the globe, and the Mammoth Steppe, while extremely cold, arid, and windy, was largely ice and snow free. The skies were clear most every day, and the sun shone down on a frigid, barren landscape of grasses, herbs, and shrubs. The steppe grasslands were surprisingly productive and produced herbs and forbs that included flowering plants like poppies, buttercups, and anemones, as well as many plants that humans eat like dandelions, sunflowers, alfalfa, parsley, and carrots.

Temperature aside, the Mammoth Steppe was not unlike the African savanna, and it was home to a wealth of herbivores that had adapted to the harsh climate and thrived on a diet of grasses and forbs. Many of these animals grew to an enormous size and are referred to as *Pleistocene megafauna*. This group included woolly mammoth, bison, yak, musk-ox, woolly rhinoceros, horse, caribou, camel, and antelope. Perhaps most impressive was the giant Irish elk, *Megaloceros giganteus*, that stood over two metres tall at the shoulder and carried antlers that spread 3.5 metres, tip to tip.

Such a wide selection of prey attracted several species of large predators. Most famous, perhaps, was the scimitar-toothed cat, *Homotherium serum*, that stood over one metre tall at the shoulder and weighed close to two hundred kilograms, the size of a modern male lion. Then there was the cave lion, *Panthera leo spelaea*, that stood roughly 1.2 metres tall and weighed in at about 360 kilograms, and a cave hyena that was larger than its modern African relative, weighing over one hundred kilograms. Other, perhaps somewhat less worrisome predators included polar, grizzly, and black bears along with various species of wolves, including the Dire wolf, all somewhat larger than their modern counterparts but not on the list of the top predators to watch out for.

The title Top Predator of the Day, however, must surely go to the short-faced bear, *Arctodus simus*, the largest known terrestrial mammalian carnivore that has ever existed. *Arctodus simus* lived in the eastern end of the Mammoth Steppe and throughout North America during the Pleistocene and weighed in at nine hundred kilograms, stood about two metres high when it was on all fours, and had a reach that exceeded four metres when it stood on its hind legs.

The top predator of the Pleistocene, *Arctodus simus*, depicted beside a six-foot Modern Human male.

Arctodus simus was a hypercarnivore and a brutal predator. Its long legs would have enabled it to run at speeds of between fifty and seventy kilometres per hour, and its massive strength meant it could eat pretty much anything it wanted, whenever it wanted. It is estimated that the bear would have required about thirty-five kilograms of meat daily just to sustain itself, and it probably consumed meat in any state it could find, whether rotting carrion or live on the hoof, paw, or foot. The chemical analysis of bones from an Alaskan specimen showed no evidence of ingestion of vegetation, however, so it seems that *Arctodus simus* wasn't interested in salads of daisies and buttercups!

During the Pleistocene glacials, which typically endured for one hundred thousand years or more, there were periods, called stadials, when temperatures suddenly dropped to levels much lower than normal, and interstadials, when temperatures spiked higher. Stadials could be severe, but they tended to be short-lived, on average lasting less than one thousand years. Interstadials typically lasted longer, up to ten millennia. They usually started quickly with a rapid rise in temperatures over the first century, followed by a slow decline that lasted for several centuries. During these relatively warm, wetter interstadials, climates moderated, rainfall increased, and green ecosystems expanded across the Mammoth Steppe.

About fifty thousand years ago, just as the early human settlers in south Asia were starting to wander northwards, the earth was entering one of these interstadial periods, and the climate of the Mammoth Steppe was moderating. Rainfall increased, and rivers and lakes filled. Forests and woodlands began to appear in the valleys of the Altai Mountains around Lake Baikal of southern Siberia and throughout Xinjiang, in western China. Even in northern Siberia, woodlands of alder, birch, and pine trees appeared.

Those that followed the Indus River Basin wandered through northern Pakistan, past the eastern entrance to the Khyber Pass, and then northeast, around the western side of the Himalayas and the Tibetan Plateau. From there, the settlers moved northeast along a series of lakes and waterways, eventually reaching the Altai Mountains of southern Russia and setting foot on the Mammoth Steppe lands for the first time, about forty-three thousand years ago.

The Altai Mountain region at the time was a mixed habitat of woodlands and grasslands with many lakes and rivers that flowed down from the glaciers in the high mountain valleys. Cold climate aside, the region was an ideal habitat with limitless hunting opportunities for the early hunter-gatherer settlers. The archaeological evidence found at several sites throughout the region indicate that, by forty thousand years ago, bands of early Modern Humans were well-adapted to life in the harsh climate conditions of the Mammoth Steppe, and they were becoming lethal predators of the megafauna that lived there.

At the base of a steep bedrock cliff overlooking the confluence of the Semisart and Kaerlyk rivers in the Altai Mountains of southern Russia, there is an open-air archaeological site known as Kara-Bom. The site was occupied by early Modern Humans for about thirty thousand years. The human remains and artifacts from different time periods have been preserved in stratified layers of sediment, enabling archaeologists to study the culture of these early humans at different time periods and understand how human culture evolved on the Mammoth Steppe from roughly forty-three to ten thousand years ago.

The oldest layer at Kara-Bom contained Mousterian industry points, scrapers, and blades that had been produced using the Levallois technique, a tradition that the first settlers had inherited from their African ancestors and still practiced some twenty thousand years later. The upper layers indicate that, over time, tool-manufacture technology had steadily advanced as sharper and finer blades and bladelets appear. This appearance may indicate the introduction of the atlatl, and possibly even the bow and arrow, into their hunting weaponry. The Kara-Bom site is rich in faunal remains, which shows that the occupants enjoyed a wide variety of high-caloric meat that included animals from the woodlands of the Altai Mountain valleys as well as megafauna from the grasslands of the steppe proper. Faunal remains at Kara-Bom include horse, woolly rhinoceros, bison, yak, antelope, hyena, and wolf, along with a variety of smaller mammals such as marmot and hare. Kara-Bom was an open-air site, and the people that lived there must have made sturdy shelters to protect themselves from the elements, particularly during the winter months. Unfortunately, no remains of these shelters have been found. They were likely made of wood frames overlaid with animal hides.

To the northeast of Kara-Bom, nestled in the foothills of the Altai Mountains of southern Siberia, is the region of Altai Krai, a land of rolling hills, grasslands, lakes, rivers, and mixed forest of conifers and broad-leaved deciduous trees. Scattered throughout the foothills are a number of limestone caves that have proven remarkably rich in artifacts and hominin remains dating to between forty-five and thirty-five thousand years ago. The caves, which include Denisova Cave, are strategically located in river valleys, close to fresh water, and they would have provided ready access to both woodland and steppe grassland game.

In addition to the evidence that both Denisovans and Neanderthals lived there, the Denisova Cave complex also contains stone tools and bone artifacts made by Modern Humans that appear to have visited the Denisova Cave about forty thousand years ago. No Modern Human remains have been found in the caves, and evidence of their presence is limited, indicating that the first Modern Human migrants into the region may not have inhabited the caves but only visited for short periods. It is possible also that the Modern Human tools were acquired by the Denisovans through trade with the new immigrants or perhaps were taught how to manufacture them on their own.

In addition to the tools, researchers have found decorative objects of bone, mammoth tusk, animal teeth, ostrich eggshell, and semiprecious stone. Because of the margin of error in dating such artifacts and hominin remains associated with them, it cannot be stated with certainty which artifacts were created by which early human subspecies. But regardless of who made them, some of the artifacts—the ostrich eggshell beads, a bone sewing needle, and a dark-green chloritolite stone bracelet, in particular—are truly objects of beauty and elegance. They indicate that their maker was a cognitively advanced, skilled artisan.

The Altai Mountain region of Siberia is archaeologically unique in that it was home to three different sapiens subspecies that coexisted in close proximity for several millennia. The Denisovans, Neanderthals, and Modern Human migrants practiced the same toolmaking tradition, hunted the same game animals, and gathered the same plants, seeds, and berries—as the seasons provided. There can be little doubt that bands of the three sapiens subspecies would have met one another, and the nature of those encounters is the subject of much speculation. The genetic evidence points to interbreeding, and it is quite likely there was inter-pairing, even mixed families. But at other times, particularly during times when game became scarce and the Siberian winters became exceptionally harsh, they would have competed, perhaps even fought, for the same resources.

But the Modern Human immigrants were more culturally advanced. They had superior technology and more effective hunting strategies, and their numbers would have quickly blossomed so that their societies and hunting bands substantially outnumbered those of the Denisovans and Neanderthals. When survival was at stake and the three peoples competed for the best habitats, resources, and food, it would have been no contest. The Modern Humans outcompeted both their rivals and went on to spread throughout southern Siberia in large numbers, while the Neanderthals and Denisovans disappeared from the region by about forty thousand years ago.

From the Atlai Mountains, the Modern Human migrants continued their wanderings eastward, fanning out into southern Siberia and northern Mongolia. Some settled along the upper Yenisei, Angara, and Lena River valleys of southern Siberia while others continued eastward, entering the forest and woodlands around Lake Baikal and the Transbaikal, between thirty-nine and thirty-five thousand years ago. By this time, the region of southern Siberia from the Altai Mountains to the Transbaikal supported a large early Modern Human population that lived in bands along the river valleys and lakes, thriving on the abundance of food resources provided by the forested regions of the valleys and the open grasslands of the Mammoth Steppe.

Much insight into the life of these early Siberians has been gained from excavations at Tobago, an open-air site located in the Khilok River valley of the Transbaikal, 230 kilometres east of Lake Baikal. The Tobago archaeological site is large, stretching some 120 metres along the riverbank and 30 metres up a gently undulating slope back from the river's edge. The site contains the remains of several dwellings and some thirty hearths. There are a number of separated activity sites that were devoted to such things as the butchering of animals, tool and weapon fabrication, animal hide-curing, and burial zones. The tool fabrication industry practiced at the site was an advanced Mousterian-Levallois tradition with more emphasis on the production of flake blades, bladelets, and points as opposed to core tools. Faunal remains at the site show that the occupants enjoyed a diverse, protein-rich diet that included large quantities of rhinoceros, horse, and sheep meat and lesser amounts of wolf, bison, reindeer, gazelle, antelope, yak, red deer, wild ass, and bear. The Tobago site was large enough to have supported a sizable population, perhaps one hundred or more, and it appears to have been home to a well-organized society that lasted for millennia.

As the population swelled in southern Siberia, bands began moving northward into central and northern Siberia. They followed the Yenisei, Angara and Lena River valleys, which snaked northward across the Mammoth Steppe through a vast, windblown, and treeless landscape of grassland and tundra, eventually emptying into the Arctic Ocean. In winter months, the sun disappeared, and the northern reaches of these rivers would have frozen solid. The entire Arctic region became a dark, empty landscape of ice and snow. But in summer months, the sun shone twenty-four hours a day, the ice melted, and the rivers flowed once again.

Game was abundant, and the people thrived, hunting the megafauna on the steppe and learning new survival strategies that allowed them to cope with the extremely harsh climate. They became skilled hunters of steppe megafauna and developed new tools and weapons to kill and butcher the larger animals.

It was at this time that prehistoric humans first became the mammoth slayers of legend.

The slaying of a mammoth would have been a cause for great celebration for people living in the Siberian Arctic. The great beasts would have provided enormous hides to be used as shelter covers, blankets, and even clothing, although smaller animal hides would have been softer and more pliable. Their tusks were used for making tools and implements as well as for carving. Bones were used as a wood substitute for constructing shelter frames over which the hides were spread. The quantity of meat was enormous, but refrigeration was readily available, and the meat could have been kept to sustain an entire band for months at a time.

Near the mouth of the Yana River, some five hundred kilometres north of the Arctic Circle, is a site where prehistoric humans made their summer camp some thirty thousand years ago. The site is near a shallow river crossing, and it may have been chosen to provide nearby hunting opportunities as animals came to the spot to cross the river. Artifacts found at the Yana site include tools made of rhinoceros horn and mammoth tusk and hundreds of stone artifacts, including choppers, scrapers, and other biface tools. Game appears to have been abundant in the area, with the most prevalent remains being those of reindeer, suggesting that the animal was their mainstay food source.

By thirty thousand years ago, Modern Humans from Africa had occupied most of the habitable regions throughout Eurasia, from the shores of the Iberian Peninsula to Northeastern Siberia, and from the Siberian arctic to the southern reaches of Tasmania.

CHAPTER TEN
The Blossoming of Human Culture

"The crucial differences which distinguish human societies and human beings are not biological. They are cultural."

...Ruth Benedict

BETWEEN FIFTY AND forty-five thousand years ago, as some bands of early humans were migrating northward from the Indian subcontinent, others were dispersing into the Persian Gulf region and the Levant. At that time, sea levels were low, and the Persian Gulf was a broad plain through which the Tigris and Euphrates Rivers meandered southward before reaching estuaries at the coast of the Arabian Sea. By about forty-six thousand years ago, the area from the eastern shores of the Mediterranean Sea to the Zagros Mountains of northern Iraq and Iran supported a large number of Modern Human hunter-gatherer societies. Band sizes grew from tribes of tens to hundreds, and with the larger numbers of people, human culture began to truly blossom.

In the region stretching from the Levant, through Anatolia and into the Balkans, there are a large number of cave sites that show evidence of human occupation from the time period fifty to forty thousand years ago. A comparative analysis of the human artifacts found in successive layers of these caves shows a transition in human culture from one based on the Mousterian-Levallois toolmaking tradition to what is known as the Aurignacian, a tradition that included advances not only in toolmaking but in other aspects of Modern Human culture as well, most particularly in artistic expression.

Aurignacian tools and weapons were much more refined and precisely made than any previous tradition. They incorporated more worked bone and antler points with grooves cut in the bottom for hafting to wooden handles and shafts. The flint tools included fine blades and bladelets struck from carefully prepared, long, prismatic stone cores as opposed to the rounder cores and flake tools produced by the older Mousterian-Levallois tradition. Generally, the blades were used as knives and scrapers and the bladelets for projectile points, but there is a much wider variety of Aurignacian tool shapes and sizes that are fit for more specific purposes.

But it is artistic achievement that truly set the Aurignacian tradition apart from the cultural traditions of earlier human societies. By thirty-five thousand years ago, the Aurignacians had become prolific artists and carvers, painting some of the earliest known cave art and carving statuettes of animal, human, and anthropomorphic figures. They had learned the fundamentals of music and fashioned musical instruments from bone and ivory. They were skilled artisans as well, fashioning pendants, bracelets, and beads from shell, wood, bone, and ivory, and they decorated their everyday utensils with graphic designs and patterns. There is evidence that they ceremonially buried their dead with elaborate ritual, and based on the nature of some of their carvings and cave art, they had a more advanced spirituality that had transitioned from animism to the more complex spiritual practice of shamanism.

Whereas prior to this time, human societies were classified in accordance with their toolmaking tradition, following the transition they were classified in terms of their cultural traditions, inclusive of toolmaking technology, but also of artistic expression, social organization, and spirituality. And the Aurignacian was the first of these cultural traditions, named after the type of site (a site considered to be the model of a particular archaeological culture), near the commune of Aurignac, in the Pyrenees of southwest France.

The appearance of the Aurignacian culture marked the beginning of a significant change in the way anthropologists and archaeologists refer to groups of ancient people. Groups of archaic humans that lived prior to this time are classified as a particular species that practiced a specific toolmaking tradition. For example, the Neanderthals practiced the Mousterian toolmaking tradition. But following the appearance of more advanced cultures, groups of people are identified and named in accordance with the cultural tradition they practiced. Hence, the term Aurignacian applies to a toolmaking tradition, a culture, and the people who practiced it. This naming tradition continues today as cultural background remains an important part of our identity as individuals.

Hence, it appears that the region of the Levant and Balkans was where our ancestors essentially completed the transition to thinking, feeling, and behaving much like we humans do today. And this transition was so abrupt, evolutionarily speaking, that it has been referred to by some as a *cultural discontinuity* or a *quantum leap in human cognition*. In the archaeological community, the transition marks the time boundary between the Middle Palaeolithic period, when human artifacts were purely utilitarian, and the Upper Palaeolithic period, when art objects appear that are undoubtedly the work of artisans with modern minds.

There is no consensus on what triggered the rather abrupt cultural revolution that occurred between fifty and thirty thousand years ago. Some scientists attribute the event to a fortuitous genetic mutation, but it is more likely that *Homo sapiens sapiens* arrived in the Levant already equipped with Modern Human cognitive capabilities, and the seemingly sudden cultural advancement was attributable to the increase in population density that occurred in the region shortly after their arrival. The area at the time was verdant and fertile, and it appears to have supported a large population of hunter-gatherers.

In any human society, modern or prehistoric, individuals are occasionally born with an unusually strong creative capacity. Throughout recorded history, the truly paradigm-shifting discoveries in science, music, art, philosophy, and other fields have typically been made by such individuals who conceive things that the rest of us are simply incapable of imagining. Such geniuses are uncommon, typically having intellectual capabilities above the ninety-nine percentile point in the statistical distribution of human creativity, but they do appear from time to time in any society. And when they do, they often produce breakthrough advancements in their chosen field.

The sudden advancements in technology and artistic expression that occurred between fifty and thirty-five thousand years ago may simply have been the result of the appearance of a few such gifted individuals, enabling them to make the first inventions in the various aspects of human culture that kick-started humankind's cultural blossoming.

But it takes more than an extraordinary mind to devise a musical scale or conceive a new technological device. It also requires a supportive and nurturing society that can tolerate the idiosyncrasies that sometimes go hand in hand with rare talent, a society that appreciates the beauty and value in the inventions of creative people and is willing to support them through patronage and moral support. Only those societies that had grown to a critical number of people, living in areas where life-sustaining resources were abundant, would have had the time and inclination to provide such an environment. With larger societies and plentiful resources, humans developed an appreciation for the aesthetic aspect of life and embraced the cultural advancements made by their eccentric geniuses. Perhaps, then, it was simply that in the Levant, societies had, for the first time, become large enough and wealthy enough to support uniquely gifted people and provide the social framework for them to devote themselves to making advancements in their chosen fields of cultural endeavour.

The incorporation of artistic endeavour into daily life is also a sign of spiritual awakening. The appearance of artistic expression indicates a dawning realization that there is more to life than survival, that life can be enjoyed, and that there is beauty in the natural world to be appreciated. Art and music enrich and add meaning to our lives by connecting us with our inner spirit. Good art resonates with our soul and amplifies our emotions, creating more intense emotional and pleasurable experiences. The arts invoke deeper emotions within us, and the more beautiful a work of art is, whether a piece of music, poetry, sculpture, or painting, the stronger the emotions it elicits. Some would say that the true measure of good artwork is the intensity of the emotional response it evokes within us.

It is our emotions that make life worth living. They are how we experience life, and without them, life would have little meaning or purpose other than to survive and procreate. That is what makes art so important and its first practice by early humans so profound. When that first human being picked up a piece of stone and began to carve it into a figurative shape, it marked the beginning of a fundamental change in human awareness, the realization that life is to be enjoyed and celebrated, not just survived. It was a major milestone in the development of human spirituality and culture.

In the Valley of the Danube

As the population in the Levant grew, human migrants began to move northwest into Anatolia and the Balkans. At the time, the Black Sea was a freshwater lake, and people and animals wandered freely between the Anatolia Peninsula and the Balkans across what is today the Bosphorus Strait.

From the Balkans, Aurignacian migrants began to disperse into western Europe, about forty-five thousand years ago. Some took the southern route, along the shores of the Aegean, Ionic, and Adriatic Seas and into present-day Greece, Italy, France, and Spain. Others headed north into the Danube River valley and followed the river westward, skirting the Alps and settling in the regions of present-day Hungary, Czechia, Austria, and Germany.

At the time the Aurignacians were entering Europe, the Danube River valley was home to several bands of Neanderthals, living in hillside caves along the valley edge or in more open settlements in the coniferous forests that grew along the valley proper. As the Aurignacians entered the region, they encountered the Neanderthals for the first time, at least in Europe, about forty-five thousand years ago.

The earliest evidence of probable interaction between the Aurignacians and Neanderthals was found in the Wachau valley of the Danube River, at the village of Willendorf, Austria. The archaeological site at Willendorf shows an abrupt transition from the Mousterian-Levallois tradition practiced by the Neanderthals to the Aurignacian toolmaking tradition about 43.5 thousand years ago.

When the Aurignacian tribes entered Europe, they greatly outnumbered the Neanderthal bands, and that, combined with their more advanced hunting weapons and techniques, resulted in the gradual marginalization of the Neanderthals. The Aurignacians took over the best dwelling places and harvested the lion's share of available resources, making life more and more difficult for the Neanderthals at a time when the climate was cold and life was already difficult. Within a millennium or two, the numbers of Neanderthals in Europe began to decline, and by forty thousand years ago, they had all but disappeared from most of Eurasia. Only a small population of Neanderthals survived the arrival of Modern Humans in central Europe by taking refuge in places such as the southern tip of the Iberian Peninsula where the Aurignacians, for some reason, had not yet arrived.

The Swabian Jura is a mountain range in southwest Germany that extends 220 kilometres from southwest to northeast and forty to seventy kilometres in width. It occupies the region bounded by the Danube River valley to the southeast, the upper Neckar River in the northwest, and in the southwest, it rises to the mountains of the Black Forest. The first discovery of the remains of *Homo heidelbergensis* was in the area of the Swabian Jura, just south of the city of Heidelberg on the Neckar River.

The geology of the Swabian Jura is mostly limestone, which formed a seabed during the Jurassic geological period. Over the millennia, rainwater seeped through cracks in the soluble limestone and formed subterranean rivers that emerge from the mountains through a large system of caves that overlook the valleys of the Ach and Lone Rivers, tributaries of the Danube. The Vogelherd, Sirgenstein, Große Grotte, Geißenklösterle, and Hohle Fels caves of the Ach and Lone River valleys of the southern Swabian Jura have been designated as a UNESCO World Heritage site because of the quantity and quality of Aurignacian cultural treasures discovered there.

The cave strata contain Mousterian-Levallois artifacts dating to as early as sixty thousand years ago and are evidence that the caves were used by Neanderthals prior to the arrival of Modern Humans. Faunal remains found from the period indicate the mainstay of the Neanderthal diet was reindeer and horse, but they also enjoyed the occasional mammoth, bovid, red deer, boar, bison, and chamois, plus small numbers of birds such as goose, ptarmigan, and grouse.

In the layers above the Neanderthal strata of the cave floors, there is an abrupt change from Mousterian to Aurignacian artifacts, dated to about forty-three thousand years ago. It would appear that the Aurignacians arrived in the Swabian Jura at that time and took up residence in the caves, essentially evicting the Neanderthal bands that had been living there when they arrived.

No Mousterian artifacts younger than about forty-three thousand years old have been found in the region, and based on the quantity of Aurignacian artifacts that replaced them, the Aurignacians must have arrived in numbers and maintained a permanent residence in the caves for several millennia. Faunal remains in the Aurignacian layers reveal a diet much like that of the Neanderthals, although they contain much larger quantities of most species, further indication that the Aurignacian population was significantly larger than the Neanderthal. There is also a sharp increase in mammoth and cave bear remains in the Aurignacian layers, supporting the assertion that the new immigrants had superior weapons and hunting strategies for killing larger and more dangerous megafauna than did the Neanderthals.

In Vogelherd Cave on the Lone River, archaeological excavations have yielded a large quantity of artifacts that have provided substantial insight into the daily life of the Aurignacians and their prolificacy as artists and artisans. The findings include a total of 217 thousand stone artifacts of various size; some twenty-eight kilograms of mammoth ivory; more than seventeen thousand tools made from bone, antler, or ivory; and over seven hundred kilograms of animal bones, including charred bones left over from cooking fires.

The discoveries contain some of the oldest examples of figurative art yet found, including 326 pierced ivory pendants and other pieces of jewellery plus several pieces of ivory that were most likely from figurines.

Of the intact pieces of art, an ivory lion's head is the oldest, dating to forty thousand years ago. It is only 5.6 centimetres long but is intricately carved with finely incised crosses on the neck. A second piece is the 3.7-centimetre-long figurine of a woolly mammoth, decorated with six short incisions and a crosshatch pattern on the soles of the pachyderm's feet. The remains of both species have been found in quantity in the cave floor, along with many pieces of carved figurines of both, suggesting that the mammoth and cave lion held a special significance to the Aurignacians that lived in the region. It is likely that the mammoth was prized for the quantity and quality of ivory, bone, hide, and meat that it provided, while the cave ion was revered for its ferocity and strength, and this reverence led the Aurignacians to feel a spiritual association with it.

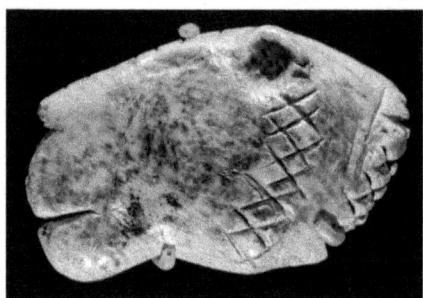 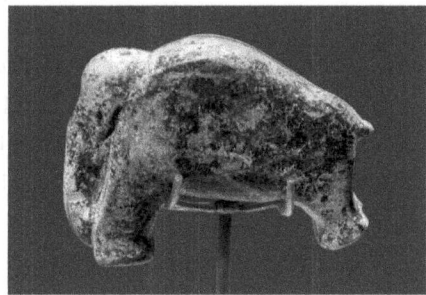

The *Lion Head* and *Woolly Mammoth* figurines found at Volgelherd Cave in the Swabian Jura.

It was also at the nearby Hohlenstein-Stadel cave site in the Lone River valley that the Löwenmensch, or Lion Man, figurine was found. The Löwenmensch has been dated to between forty and thirty-five thousand years ago, providing strong evidence that the Aurignacians had developed an animistic belief system and practiced shamanism. And if so, they would have been the first society known to have done so. The fact that the cave lion was selected for the head of the anthropomorphic piece also reinforces the notion that the Aurignacians held a special spiritual association with the animal.

In 2008, a small, six-centimetre-long, mammoth ivory carving of the human female form was found at Hohle Fels Cave in the nearby Ach River valley. The figurine, known as the *Venus of Hohle Fels*, has been dated at between forty and thirty-five thousand years ago.

The *Venus of Hohle Fels*.

The *Venus of Hohle Fels* carving is only one of many statuettes of the female form made during the Palaeolithic period throughout Europe, and archaeologists have coined the term *Venus figurines* for the collection of over two hundred such statues that have been found. Palaeolithic Venus figurines were carved from all manner of materials, ranging from soft stone (such as clay, steatite, calcite, or limestone) to bone and ivory. And while produced at sites widely separated in both time and distance, all Venus figurines exhibit strikingly similar physical attributes, including curvaceous bodies with large breasts, bottoms, abdomen, hips, and thighs. They are usually tapered to the head and feet. Archaeologists have advanced many theories about the meaning or purpose of the figurines, ranging from serving as a ritual or symbolic spiritual function to prehistoric erotica, but no archaeological evidence has been found to support any of the proposed theories. But given that the Aurignacians and later cultures were skilled artisans, it is likely that the Venus figurines were deliberately carved with exaggerated proportions in order to emphasize those aspects of the female figure that the artists considered attractive, for whatever reason—sexual, aesthetic, or perhaps related to fecundity.

In addition to the figurines found at Swabian Jura, researchers discovered bone and ivory segments of flutes, dating to forty-two thousand years ago, at Geißenklösterle Cave. Some of the ivory flutes were made from woolly mammoth tusk and would have required considerable skill to craft. First, the ivory had to be carved into a long, thin rod that was then split lengthwise into two, half-cylindrical pieces. Each of the pieces was then hollowed out, and the air holes were drilled out on one of the pieces. The two pieces were then reassembled with glue to make an airtight cylinder. Seems easy enough until you remember that this was all done with tools of flint or obsidian and without modern glue!

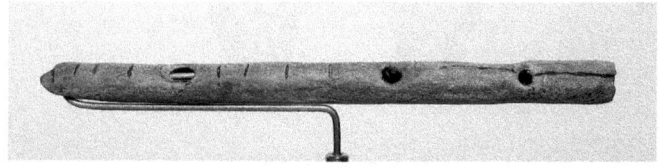

A bone flute found at Geißenklösterle Cave in the Swabian Jura.

At the Hohle Fels cave site, enough matching flute fragments were found to reassemble them into a nearly complete instrument, which turned out to be a five-holed, pentatonic flute, dated to approximately thirty-five thousand years ago. The Hohle Fels flute had been fashioned from the naturally hollow wing bone of a griffon vulture. It is long and slender, only eight millimetres in diameter and thirty-four centimetres long.

The Hohle Fels flute was too delicate to play, so scientists made an exact replica, reverse-engineered from measurements of the original; they found it to be quite comparable to a modern-day flute. When played, the replica produced a complete pentatonic scale, one of the world's most commonly used scales even today, with near-perfect pitch.

The construction of a pentatonic flute with perfect pitch is an acquired skill even for modern instrument makers with computer design algorithms and precision electric tools. How did someone who had no prior knowledge of wind instruments and only flint or obsidian tools with which to craft it manage? The prehistoric flute maker's knowledge of where to place the holes to produce accurate pitch could only have been the result of significant trial and error, guided by what sounded harmonious. The process would have involved the development of countless prototypes, perhaps even requiring cumulative knowledge passed down through generations of musicians and artisans. Careful inspection of the artifact shows that small lines were first etched into the flute, indicating where to drill the holes, a sign that by this point in time, the craftsman had acquired the musical knowledge of where the holes had to be in order to produce melodic sounds. That knowledge would have been stored on a physical device, perhaps on the world's first ruler!

The Hohle Fels flute would have been technically capable of playing contemporary music, a testimony to not only the talent of the artisan that made it, but to the universal and timeless appeal of music itself. One can only wonder about the first sounds, the first melodies that would have been played on the Swabian Jura flutes. Would these early artisans be musical composers as well? Did they hear melodies in their minds the way modern composers and musicians do, and would they have learned to play those melodies on their newly fashioned flutes? Or were the Aurignacians already singing songs and fashioned the flutes to accompany those melodies?

Downstream from the Swabian Jura on the Danube River, lies the modern-day city of Salzburg, birthplace of Wolfgang Amadeus Mozart, one of the greatest musical composers to have ever lived. In 1778, Mozart composed his *Flute Concerto No. 1 in G major*, a masterpiece that is still played in concert halls along the Danube River valley. And so it would seem that, from the Hohle Fels cave of the Swabian Jura to the concert halls of Vienna, the sweet sound of the flute has wafted through the valley of the Danube for over forty thousand years.

The astounding collection of Aurignacian artifacts found at the Lone and Ach River valley cave sites indicates that there was a thriving society of people living in the region for many millennia. They were hunter-gatherers, so they would have lived in small, somewhat separated bands or family groups to avoid exhausting the resources in the region, but still close enough to allow for communication and cooperation. It is likely that the various groups socialized; shared resources, knowledge, art, and music; and inter-partnered. The picture that emerges is of a community of several hundred people living out their lives along the Danube River valley, going about their daily routines, united by the bonds of family, friendship, and a shared spirituality. Under such conditions, cultural advancement typically takes place rapidly, which may explain why so many elegant works of art and artisanship have been found at the Swabian Jura cave sites.

Into Europe by the Southern Route

About the same time that bands of Aurignacians were entering the Danube River valley, others were migrating westward, along the northern shores of the eastern Mediterranean Sea. At the time, sea levels were low, and many of the islands of the Aegean archipelago were connected by land bridges and coastal plains, linking them with Anatolia and the Peloponnese, while the North Adriatic Basin was a broad, coastal plain that extended some distance south into the Adriatic Sea.

At the time the Aurignacians arrived in the region, approximately forty-five thousand years ago, there was a large Neanderthal population living at many sites spread along the coastlines of the Aegean, Ionian, Adriatic, and Tyrrhenian Seas. And as the Aurignacians dispersed into the area, the archaeological record indicates that, in sharp contrast to what appears to have happened in the northern regions, the encounters between the two peoples were rather amicable. A complex social interaction took place that included cultural exchanges, integration, and most likely, the formation of mixed families and social groups.

The blending of the two cultures gave rise to a transitional culture, known as Uluzzian, characterized by influences from both the Mousterian *lithic-core-based* tradition of the Neanderthals and the more refined, *lithic-flake-based* tradition of the Aurignacians. While more refined and prolific than the Mousterian, the Uluzzian culture was still primarily utilitarian in nature and no artistic artifacts, such as jewellery or figurative art, have been found. Evidence of the Uluzzian culture has been found at dozens of cave and open-air archaeological sites around the Aegean as well as along the Adriatic and Tyrrhenian coastlines of peninsular Italy. The term Uluzzian is derived from the location of its first discovery, the Grotto del Cavallo located at Uluzzo Bay at the southern end of Italy. Grotto del Cavallo is regarded as the type site for the Uluzzian culture.

It is perhaps puzzling that the encounters between the Neanderthals and Aurignacians along the Mediterranean coast appear to have been rather amicable, whereas the encounters in the Swabian Jura and other northern sites appear markedly less so. In the southern encounters, the archaeological evidence shows that the two subspecies coexisted in integrated societies for thousands of years, but the evidence in the north shows a sharp, stratigraphic changeover from one culture to the other, suggesting the Aurignacians displaced the northern Neanderthals in relatively short order whenever encounters took place.

Perhaps the explanation lies in the fact that in the Swabian Jura and other northerly regions, the climate was much harsher and life-sustaining resources were much more difficult to acquire than in the south. With the arrival of the Aurignacians in the Danube River valley, there would not have been enough shelters and resources to accommodate both the Neanderthals and the large number of new arrivals, so the two groups would have been forced to compete. The Aurignacians, with better technology and far greater numbers, would have easily taken over the best dwelling places and hunting areas, forcing the Neanderthals to emigrate into less hospitable areas. Under such circumstances, there would have been little or no social interaction between the two peoples. And as the Aurignacians occupied more and more of the habitable northern and central European areas, the Neanderthals were driven out into ever less hospitable regions until their numbers gradually diminished into extinction.

In the eastern Mediterranean region, however, the climate was more benign, and the bounty of the sea would have provided enough for all, reducing the need to compete for resources. Under such benign conditions, coexistence and social integration would have been beneficial to both peoples, and that appears to have been what happened.

And so, while the eventual disappearance of the Neanderthals from the northerly regions appears to have been mostly due to displacement and marginalization, in the south, it likely had more to do with social integration and genetic absorption. Given that the population of Aurignacians was much greater than that of the Neanderthals, the genetic absorption would have occurred over relatively few generations. This supposition explains both the disappearance of the Neanderthals by about forty thousand years ago and the relatively small Neanderthal contribution to the Modern Human genome observed today.

The record of the Uluzzian culture first appears in eastern Mediterranean archaeological sites dating about forty-five thousand years ago, and the culture continued to flourish throughout the region for over five thousand years. But by approximately thirty-nine thousand years ago, the Uluzzian people seem to have virtually disappeared as the number of artifacts contained in the stratigraphic layers of most cave sites in the region are essentially nonexistent after this time.

The cause of the sudden disappearance of the Uluzzians remains the subject of some debate amongst archaeologists. Many attribute the disappearance to the eruption of the Archiflegreo volcano—the thirteen-kilometre-wide caldera known today as the Phlegraean Fields. It is located on the Campanian Plain in southwest Italy and is one of the most complex volcanic structures in the world. Over the millennia, several massive volcanic eruptions have occurred in Campania with drastic consequences for the flora and fauna of the region, the most famous of which occurred in 79 CE when Mount Vesuvius smothered the region in volcanic ash, completely destroying the Roman city of Pompeii and killing most of its inhabitants.

The most recent estimates for the Archiflegreo eruption, known as the Campanian Ignimbrite Supereruption, or CI eruption, dates it at approximately 39.28 thousand years ago. Studies have revealed that the eruption released massive quantities of volcanic ash, pumice, and lithic debris into the atmosphere. The debris then dispersed and settled over an area of almost four million square kilometres, stretching from southern Italy across the eastern Mediterranean landscape as far east as central Russia and the Caspian Sea. The studies have also found geographical evidence that points to major disruptions in climate and an alteration of the evolutionary direction of much of the flora and fauna in the region following the event.

At the time of the CI eruption, the last great ice age of the Pleistocene (known in North America as the Wisconsin and in northern Europe as the Weichselian) was advancing from the north, and massive icefields covered

Scandinavia, the British Isles, and much of northern France, Belgium, the Netherlands, and Germany. At the southern edge of the ice sheets, extending from western Europe through Siberia and into Alaska, was the cold grassland and tundra plains of the Mammoth Steppe. In the south of France and Spain, the forests and woodlands shrank into river valleys and the lower mountain slopes.

Exacerbating these worsening conditions, the CI eruption triggered what is known as a *Heinrich event*, which caused global temperatures to suddenly drop further by 1 to 2°C for a period of two to three years. The event accelerated the onset of the Great Ice Age, and living conditions throughout Eurasia for the Neanderthals and early Modern Humans became even more harsh and challenging. In fact, the impact of the CI eruption was so devastating that archaeologists consider it to have been a significant factor in the disappearance of the Neanderthals, which was essentially complete throughout Eurasia shortly after the eruption occurred.

By about thirty-five thousand years ago, bands of Aurignacians were entering southwest Europe, taking up residence in the river valleys of southern France and the northern coastal regions of Spain. In these regions, known as ice age refugia, conditions, while colder, were not unlike those found today with forested valleys and lowlands, providing a viable habitat for animals and early humans seeking refuge during the harshest periods of the ice age. For humans that had adapted to life in cold climates, the river valleys proved ideal places to live. They provided shelter from the elements, plenty of woodland game, water to drink, and access to megafauna on the Mammoth Steppe.

Principal amongst these ice age refugia were the Ardèche River valley of southeast France; the Vezere River valley in the department (province) of Dordogne, France; and the province of Cantabria in northern Spain. Archaeological evidence found at these sites indicates largely continuous occupation by early humans from approximately thirty-five thousand years ago to the end of the Great Ice Age, around twelve thousand years ago, by which time the earth's average annual temperature had risen to the levels experienced in modern times.

The Ardèche River is a small tributary of the Rhone in south-central France. The source of the river lies 1,467 metres above sea level near the

Col de la Chavade, in the forest of Mazan and the commune of Astete. The valley of the Ardèche is very scenic, in particular, along a thirty-kilometre section known as the Ardèche Gorges where limestone cliffs tower up to three hundred metres above the river. From the gorges, the river winds its way through the orchards and vineyards of the lower plateau, eventually reaching the Rhone River at Pont Saint Esprit.

The Ardèche's most well-known feature is a natural, sixty-six-metre-high stone arch that spans the river and is known as the Pont d'Arc. It was formed when the river broke through a narrow escarpment between two of its meanders, only a few thousand years ago, long after the area was settled by the Aurignacians, thirty-seven thousand years ago.

The gorges of the Ardèche are the site of numerous caves, most significant of which is the Chauvet, situated just upstream and along the course the river followed before the breach that formed the Pont d'Arc. The entire cave is about five hundred metres long and consists of numerous chambers connected by passages. Most of the rooms are filled with stalactites and stalagmites interspersed with curtain-like sheets of sparkling limestone. The soft, clay-like floor of the cave retains the paw prints of cave bear along with large, rounded depressions that are believed to be the places where the bears slept during hibernation. Fossilized bones are abundant and include the skulls of cave bears, the horned skull of an ibex, and the remains of wolves. The cave bears also left innumerable scratches on the walls.

The walls of the caverns and passageways are alive with several hundred animal paintings that depict many of the fauna that lived in the area at the time. There are panels depicting horses, mammoths, cattle, bison, rhinoceros, ibex, reindeer, deer, auroch, musk-oxen, panther, and owls, along with many predators including cave lions, panthers, bears, and cave hyenas. There are also a few panels of red ochre hand stencils, along with lines and dots. The most impressive artwork, done mainly with black pigment, has been dated to the earlier period and has been attributed to the Aurignacians.

Dating the art in the caves has proven problematic, and several attempts have been made with widely varying results. A study published in 2012 supports placing the art in the Aurignacian period, approximately thirty-two to thirty thousand years ago, but a more recent study, published in

2016, that used an additional eighty-eight radiocarbon dates, showed two periods of habitation, one between 37 and 33.5 thousand years ago and the second between 31 and 28 thousand years ago. These results place the first occupation during the Aurignacian period and the second in the Gravettian. Between the occupation periods, the cave appears to have been visited only by cave bear and other animals.

The artists who produced the paintings used techniques rarely found in other cave art. Many of the paintings appear to have been made only after the walls were scraped clear of debris and concretions, which would leave a smoother and noticeably lighter area upon which the artists worked. Similarly, a three-dimensional quality and the suggestion of movement are achieved by etching around the outlines of certain figures. The art is also exceptional for its time because it depicts scenes of animals interacting with each other—one panel shows a pair of woolly rhinoceroses butting horns.

Of all the art in the Chauvet Cave, two panels stand out as demonstrating the astonishing technical skill and artistic talent of the Aurignacian artists. The first is the *Fighting Rhino and Horses* panel, which contains several animals in a dramatic setting. There are two rhinoceroses confronting each other, the heads of four galloping horses that appear nervous as if attempting to escape predators, two additional rhinoceroses, a stag, and two woolly mammoths. The artist who drew the horses mixed charcoal and clay to obtain various hues and visual effects such as shading and perspective. The technique used was primarily stump-drawing, as well as scraping of the outer edges of the images to highlight them with a pale aura.

A second, equally impressive panel, known as *The Panel of the Lions*, contains a pride of hunting lions, and the expressions on the faces of the lions are clearly those of predators stalking prey. The two panels are such that some have suggested the panels were deliberately placed in close proximity so as to create the effect of a group of nervous, hard-breathing horses being intently stalked by the pride of lions.

Replicas of two Aurignacian cave paintings from the Chauvet Cave, the *Fighting Rhino and Horses* Panel (left) and *The Panel of the Lions* (right).

Careful exploration of Chauvet has shown that, although cave bears and other animals had made extensive use of the cave, humans do not appear to have occupied the cave as their principal residence and probably only entered it on special occasions. Many of the paintings were done in the deeper recesses where there was little or no natural light, and the artists and visitors would have brought torches with them to create or admire the paintings. The implication is that the paintings were not intended as everyday artwork located in living areas for all to enjoy. Rather, they must have had a special, perhaps spiritual, significance, and were located in a place reserved for ceremonies or ritual practices of some kind, much like the ornate cathedrals and mosques of modern times.

Another intriguing find at Chauvet, dated to the Gravettian occupation, are two sets of footprints, one left by a child and the other by a large wolf. The footprints do not overlap but look as if the two were walking side by side, and they extend well into the darker parts of the cave. It has been determined that the prints were very likely made at the same time, and there is no indication that the wolf was preying upon the child. The tracks indicate that both the child and the wolf were walking at a normal, unhurried pace. The fact that the tracks never cross also supports the theory that the two were walking side by side. If they had been there at separate times or if the wolf had been stalking the child, the tracks would have at least occasionally overlapped, especially at very narrow points in the path.

Some paleontologists speculate, therefore, that the child and the wolf were walking into the cave together and that the wolf was a pet, possibly nurtured by the child from when it was a young, orphaned cub. The truth behind the tracks can never be known, but the image of a prehistoric child and his or her pet wolf walking into Chauvet cave, perhaps to visit with a parent or elder that was painting by torchlight, is so enchanting that one can only hope that no evidence will ever be found to dispel it.

Twenty thousand years ago, a landslide blocked the entrance to Chauvet Cave, preserving its precious heritage until its discovery in 1994, and in 2014, Chauvet-Pont-d'Arc Cave was declared a UNESCO World Heritage site in order to protect the cultural treasures it contains.

The Gravettians

The Aurignacian people thrived throughout southern Europe and increased in numbers, occupying most of the cave sites in the refugia areas. Their culture advanced over time as their tools became ever more refined and specialized and their weapons more deadly. And while cultural changes and advances were typically gradual and more or less continuous throughout palaeolithic Europe, paleoanthropologists nevertheless identify new cultural traditions when significant changes in toolmaking or artistic endeavour are observed. In accordance with this tradition, the Aurignacian culture is deemed to have largely disappeared by about twenty-nine thousand years ago when a new cultural tradition, known as the *Gravettian*, began to emerge throughout Europe.

The Gravettian culture is named after the type site of La Gravette in the Dordogne region of France where its characteristic tools were first found and studied. The Gravettian period lasted from thirty-three until approximately twenty-two thousand years ago, which was a particularly cold period throughout most of Europe, and the harsh climate conditions shaped both the Gravettian lifestyle and culture.

The Gravettians were lethal hunters, and their diet consisted mostly of meat, which provided the high-caloric intake required to sustain them in the frigid environments in which they lived. Their culture and lifestyle centred on hunting animals of all sizes, and they adopted a mobile, complex hunting lifestyle that allowed them to migrate with their food sources, or they built their encampments at strategic sites such as river fords and narrow passageways in valleys where animals like deer and reindeer were forced to pass during their migrations.

The Gravettian diet included larger animals such as mammoth, deer, hyena, wolf, and reindeer, which they killed with stone or bone weapons, and hares and foxes captured with woven nets. They were an innovative people, fashioning stone oil lamps and creating figurines and works of art from fired clay. Some archaeologists speculate that they may have also invented the atlatl and possibly even the bow for thrusting spears and shooting arrows with amplified force.

Gravettian tools and weapons were characterized by tanged points known as Gravette Points, and their blades were backed or blunted on one edge to facilitate handling and hafting. In addition to producing more refined hunting and butchering points and tools, they added ornamentation to tools and weapons and carved works of art from wood, bone, ivory, and clay.

Like the Aurignacians before them, the Gravettians were prolific artists. Surviving Gravettian art includes numerous cave paintings, objects of jewellery, and especially small, portable Venus figurines made from ivory and fired clay. The Gravettian Venus figurines generally conform to the same, specific physical type that characterized such carvings throughout Europe during the Palaeolithic in that they all have large breasts, broad hips, and prominent posteriors. The statuettes tend to lack facial details and are typically tapered at the head and feet.

The Gravettians were widespread throughout Europe. Their range extended from Portugal, through Spain, France, Italy, and Britain in the west, and through Eastern Europe to Russia in the east. In the west, they lived primarily in caves or semi-subterranean dwellings, while in the east, they lived in villages of multiple, rounded dwellings of wood, mammoth bone, and hides, typically arranged in small villages.

Between approximately twenty-nine and twenty-five thousand years ago, a variant or subculture of the Gravettian, known as *Pavlonian*, appeared in an eastern region that included Moravia, northern Austria and southern Poland. Its name is derived from the village of Pavlov, located in the Pavlov Hills of southern Moravia where evidence of the culture was first found.

The Pavlovians lived on the tundra at the fringe of the ice sheets that extended south from Scandinavia at the time, and their culture was focused on the hunting of herds of mammoth that were found there. The great beasts provided meat, fat fuel, hides for tents, and large bones and tusks for framing shelters.

Artifacts found at Pavlonian sites include a variety of flint implements, polished and drilled stone artifacts, bone spearheads, needles, digging tools, flutes, bone ornaments, drilled animal teeth, and seashells. Artistic carvings and figurines of humans and animals made of mammoth tusk, stone, and fired clay have also been found, and a textile impression made in wet clay provides the oldest evidence of weaving by humans ever found.

Along a small stream, near Dolni Vestonice in the Czech Republic, there is a Pavlonian open-air archaeological site dated to twenty-six thousand years ago. The site is one of the oldest permanent human settlements ever found by archaeologists, anywhere.

The people that lived at Dolni Vestonice were prolific steppe hunters, and the site contains large quantities of mammoth and other large animal bones that were used to construct a fence-like boundary around the settlement. At the centre of the enclosure were several hearths, surrounded by substantial, well-constructed dwellings as evidenced by postholes and stone foundations. The dwellings were likely made with frames of wooden poles or mammoth bone, over which large coverings of mammoth or other animal hides were laid, tightly sewn together to provide protection from the elements.

One of the most significant discoveries at the site was figurines and art objects made of fired and unfired clay, the earliest evidence of ceramic pottery yet found. The pottery appears to have been made at a kiln located about eighty metres upstream of the main living enclosure in a lean-to shelter dug into an embankment. An estimated 2,300 ceramic artifacts, including figurines of bear, lion, mammoth, horse, fox, rhinoceros, and owl as well as two Venus figurines, were found associated with the kiln. One of the female figurines, known as the *Black Venus*, stands about eleven centimetres tall and is the earliest known ceramic figurine of the human female form.

It appears that the Pavlonians used ceramics exclusively for art as no shards of pottery vessels or containers have been found. This may be because the ceramics produced at Dolni Vestonice had all been fired at too low a heat to make them durable enough for everyday use.

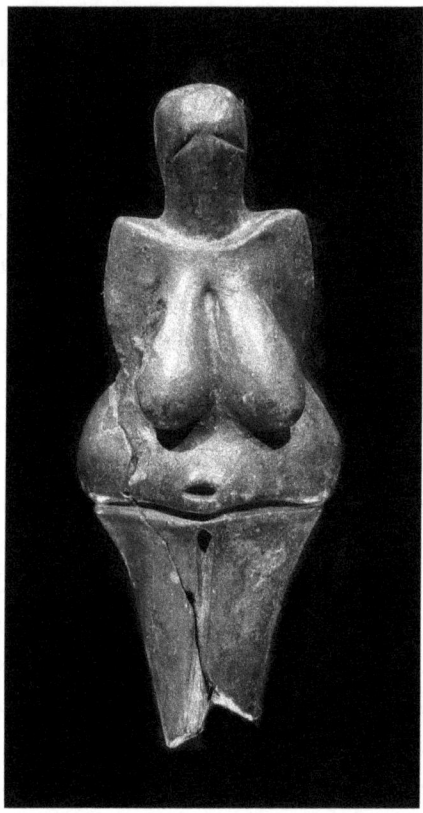

The *Black Venus of Dolni Vestonice*.

In addition to ceramic art, the Gravettians carved artistic items from mammoth ivory, including Venus and animal figurines, engravings, and articles of personal adornment. An intriguing carved ivory figure in the shape of a female head with a distorted left side of her face was discovered near the shelters inside the enclosure. In a nearby burial, a female skeleton was unearthed, with the left side of her skull disfigured in the same manner as the carved ivory figure, indicating that the figure was an intentional depiction of this specific individual. The woman, just over forty years of age at death, had been ritualistically placed beneath a pair of mammoth scapulae, one leaning against the other. The bones and the earth surrounding the body contained traces of red ochre, a flint spearhead had been placed near the skull, and one hand held the body of a fox. Archaeologists interpret this as evidence that the woman was a shaman.

Another interesting artifact found at Dolni Vestonice is a wolf bone that is marked with fifty-five grooves and may have been used for tallying or, possibly, measuring. The bone has been dated to thirty thousand years ago, which places it near the beginning of the Gravettian period.

The imprints of textiles found at the site suggest that the Gravettians were weavers as well and used a variety of techniques to make baskets and nets for hunting or fishing. Woven cloth and shells found at the site have been shown to originate from the Mediterranean coast, so that the people of Dolni Vestonice either travelled to collect them or traded for them with other tribes that passed through the area.

The Solutreans

At the end of the Gravettian period, Europe was entering the grip of the *Last Glacial Maximum*, a period of about four thousand years when the earth's temperature plummeted to the lowest temperatures it had experienced within the preceding five hundred million years.

During the Last Glacial Maximum, icefields covered much of northern Europe, and conditions on the Mammoth Steppe grew extremely harsh, presenting survival challenges to both man and beast. People throughout the European continent were forced to move ever more southward, ahead of the advancing glacial and into the refugia of southern France and Spain. In these areas, bands and tribes of early humans thrived in spite of the harshness of the climate, and a new, more advanced culture emerged known as the Solutrean.

The Solutrean culture thrived between approximately twenty-two and seventeen thousand years ago and is named after the type site at the Rock of Solutry, a limestone escarpment eight kilometres west of Macon in the south of the Bourgogne in France.

Solutrean toolmaking employed techniques that had not been seen before; they produced finely worked, bifacial points knapped with antler and hardwood batons. This method permitted the working of delicate slivers of flint spearheads and barbed and tanged arrowheads, chisels with a sharp edge on one end, finely crafted flint knives, and saws. Long spear points with a tang and shoulder on one side only are also characteristic implements of this industry. The hallmark of the Solutrean tool industry were bifaced, laurel-leaf points that were used as both tools and weapons. Some of the smaller bifaces were hafted onto the ends of spears, darts, or arrows, and the larger laurel-leaf points were hafted onto short handles to make knives.

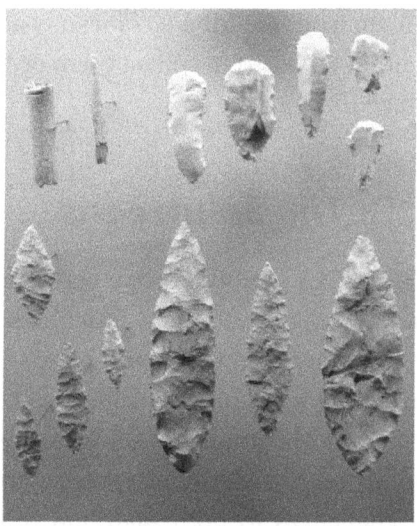

Solutrean laurel leaf-points and tools.

The first definitive evidence of the use of the bow and arrow has been associated with the Solutreans. Cave paintings in the Cova dels Cavalls at Valltorta Gorge in Spain, dating to twenty-two thousand years ago, show hunters with bows and arrows aiming at a herd of deer. Some of the animals are depicted with arrows protruding from their bodies.

The oldest definitive evidence of the use of the bow and arrow from the Cova dels Cavalls in Spain.

The Solutrean lithic technology is similar to the Clovis points that appeared throughout America about thirteen thousand years ago, and some archaeologists have speculated that the close resemblance is no coincidence. According to what has become known as the *Solutrean hypothesis*, people of the Solutrean culture in Europe migrated to North America by boat along the pack ice of the North Atlantic Ocean. They survived by using skills similar to those of the modern Inuit people: hauling out on ice flows at night, collecting fresh water by melting ice, hunting seals and fish for food, and using seal blubber as heating and cooking fuel.

The theory speculates that the Solutrean voyagers brought their toolmaking tradition with them, which formed the basis for the Clovis technology. Recent archaeological discoveries of what might be interpreted as a transitional technology between the Solutrean and what later became Clovis technology at Meadowcroft Rockshelter in Pennsylvania, Cactus Hill in Virginia, and Miles Point in Maryland lend some credence to the theory.

However, since the only evidence to support the Solutrean hypothesis are the similarities between the Clovis and Solutrean technologies, which may simply be coincidental, few archaeologists today support the Solutrean hypothesis. Furthermore, recent genetic studies that compare the DNA of extant Native Americans with east Asian populations strongly point to an Asian origin for the indigenous people of the Americas.

A few archaeologists, however, continue to actively work various archaeological sites in the eastern United States in search of more conclusive, supportive evidence of the hypothesis. And while the majority of the indigenous people of the Americas are likely descendants of immigrants from eastern Siberia, the possibility that Solutreans wandered as far as the Americas and established settlements in the New England area, at least for a period of time, cannot be entirely ruled out. Definitive evidence of localized Solutrean populations in the eastern United States may yet be discovered.

The Magdalenians

Following the Last Glacial Maximum, the earth's temperature began to rise sharply, geologically speaking, and climate throughout Europe began to return to warmer, less arid conditions. The harsh environment of the Mammoth Steppe moderated, and the large megafauna began to move north once again, following the retreating ice sheets as they slowly melted.

It was at this time, starting about seventeen thousand years ago and continuing to twelve thousand years ago, that the Magdalenian culture began to appear in Spain and the south of France. Within a millennium or two, bands of Magdalenians began to leave the southern refugia to follow the retreating ice, dispersing northward onto the Mammoth Steppe. Being hunter-gatherers, the Magdalenians were quick to seize upon the bountiful hunting and foraging opportunities they found there, and they followed herds of herbivores on their seasonal migrations, which took them across the steppes into most regions of northern and central Europe. Before long, the Magdalenians had settled throughout much of the European continent, from Portugal in the west to Poland in the east.

The Magdalenian culture is named after the type site of La Madeleine, a rock shelter located in the Vézère valley, in the Dordogne department of France. The Magdalenian period lasted some six thousand years, and their culture continued to evolve and advance throughout that time period. Their tool industry was characterized by a transition away from stone knapping as their primary production process to carving bone and ivory for most of their implements and weapons. This transition had been started by the Solutreans, but by the end of the Magdalenians' tenure, most of their tools and weapons were made of bone, ivory, and antler, prompting some archaeologists to refer to the later Magdalenian period as the Bone Age. The bone instruments became ever more specialized and ornate and included new designs for spear and arrowheads, harpoon heads, borers, hooks, and needles.

Magdalenian tools and harpoon point.

A Magdalenian carving of a bison apparently licking an insect bite.

While flint was still used for jobs requiring harder, heavier tools, the Magdalenians are best known for their elaborate worked bone, antler, and ivory that served both functional and aesthetic purposes. It is interesting to note that in one reindeer antler etching, a man is portrayed as naked, which suggests a warm climate. There is also evidence that the fauna of the period included tigers and other tropical species side by side with reindeer, foxes, Arctic hare, and other polar creatures. The Magdalenian period truly was a time of cultural and environmental transition.

Another accomplishment of the Magdalenians for which we Modern Humans are deeply indebted was the domestication of the dog. All dogs in the world today are descended from a now-extinct species of wolf that existed throughout Eurasia in prehistoric times. It is likely that these wolves were domesticated at many parts of Eurasia around the same time, twelve to fifteen thousand years ago. There are various theories of how this domestication occurred, one being that packs of wolves started visiting Magdalenian campsites to feast on the bones and other remains that the people had discarded and, over time, came to depend on the habit as a principal source of food. Eventually, a mutual trust would have developed between the wolves and humans. Humans began to actively feed the animals, and the animals began to stay in the vicinity of the camp.

As successive generations appeared, the wolves became dependent on the humans for food, and the humans, in turn, became dependent on the wolves for companionship and security.

Another theory is that humans adopted orphaned wolf cubs as pets and raised them to be part of the human social group. Subsequent generations of these pet wolves would have known no other life and would have stayed with the human societies as a result.

Less than three hundred kilometres to the west of Chauvet-Pont-d'Arc, lies the Vezere River valley in the department of Dordogne, also in southern France. The valley contains almost 150 prehistoric sites, mostly dating to the Magdalenian period. There are twenty-five caves in the valley with extensive murals and wall art, many of them astonishing in their grandeur and beauty. The natural colours and rugged formations of the caves serve only to further highlight the magnificence of the art, which transforms the caves into breathtaking prehistoric art galleries.

One of the most famous of the Vezere valley caves is Lascaux, located near the village of Montignac. The Lascaux cave extends for 240 metres underground and has many galleries and passageways that have been named after the paintings found within them. The cave as a whole contains almost six hundred paintings of horses, deer, aurochs, ibex, bison, and even large cats arranged in hunting scenes and other compositions. Each animal is drawn with remarkable lifelike detail and painted with rich colours. In addition to the paintings, there are about fourteen hundred engravings of similar subjects and quality.

Lascaux's most famous chambers include the *Hall of the Bulls*, the *Axial Gallery*, the *Apse*, and the *Shaft* that together contain some two thousand images, about nine hundred of which are animals; the remainder are geometric symbols of varying shapes.

The *Hall of the Bulls* at Lascaux.

The sheer number of images, their size and exceptional realism, as well as their spectacular colours explain why Lascaux is sometimes referred to as the Sistine Chapel of Prehistory. Not long after its opening in 1948, Pablo Picasso paid a visit and was amazed at the quality of the cave's art, saying that as artists, man had learned nothing new since the cave art was produced.

In 1979, Lascaux was added to the list of UNESCO World Heritage sites, together with another 147 prehistoric sites and twenty-five decorated caves located in the Vezere valley of the Correze and Dordogne regions. The art of the Vezere valley falls in the middle of the Magdalenian period and is a testimony to the exceptional talent of the artists and cultural achievements of the Magdalenian people. It is clear from their toolmaking industry and artistic achievements that the Magdalenians were cognitively, emotionally, and behaviourally as modern as any of us today.

At the end of the Last Glacial Maximum, the vast northern ice sheets began to retreat as the earth entered what is referred to as the Holocene epoch. The earth's global mean temperature stabilized and remained within a relatively narrow range from year to year into modern times. The climate throughout Eurasia became more temperate and predictable. And with the moderating climate, the Magdalenians began to migrate north from their refugia in southern France and Spain, and by twelve thousand years ago, there were large populations of people living in virtually all regions of the European continent. New cultures were appearing as language, technology, artistic expression, and social customs began to evolve in isolation, laying the foundation for the many different cultures found across the European continent today.

Similarly, advanced cultures began to emerge in different parts of the world. Today, some forty-five thousand years after the appearance of the first Aurignacian culture, there are hundreds, if not thousands, of societies throughout the earth, each with its own unique culture. A society's culture is strongly influenced by the environment in which it finds itself, and geographically separate bands or tribes of early humans naturally developed different cultures. Different geographical environments present different opportunities for acquiring scientific knowledge, demand different technologies and survival behaviour, and create the need for different inventions. *Homo sapiens* tribes in Eurasia lived a very different lifestyle than those in Africa or Asia. They had to cope with a much different climate with different degrees of seasonal and cyclical climate change. They hunted different game; foraged for different fruits, nuts, and vegetables; and built different dwellings. It is not surprising that by the dawn of recorded history, there was a large variety of cultures on earth, each with different language, social organization, scientific knowledge, technology, and artistic expression.

As the climate became more moderate and reliable, other cultural changes began to appear throughout Eurasia. Starting some twelve thousand years ago, people in the fertile crescent of the Levant were beginning to experiment with the cultivation of wild forms of wheat, barley, lentils and peas, genetically modifying them through selective reseeding into today's grains and vegetables. Simultaneously, wild boar, sheep and auroch were being domesticated into pigs, sheep, and cattle. Meanwhile, in faraway China, along the banks of the Huang He, rice, soya, and azuki beans were being planted and harvested.

A profound change to the way people lived their lives was beginning, and within a few millennia, the hunter-gatherer way of life would give way to the more sedentary lifestyle of agrarian societies in a few parts of the world. The seeds of civilization were being sown, and the world would never be the same again.

CHAPTER ELEVEN
Peopling of the Americas

BOTH PALEONTOLOGICAL EVIDENCE and recent genetic research have left little doubt that the indigenous people of the Americas are descended from a basal ancestral lineage that formed in Siberia prior to the Last Glacial Maximum, between thirty-six and twenty-five thousand years ago. Subsequently, starting about twenty thousand years ago, they began their northeasterly migrations that would lead them into the Americas.

As the last Great Ice Age was locking much of northern Eurasia in its icy grip, not all of the people in the northern regions migrated south to seek refuge in the more hospitable climate of southern Europe and central Siberia. Some headed northeast into the region known as Beringia.

During the period from about thirty to ten thousand years ago, sea levels were so low that the Bering Strait and parts of the adjoining Chukchi and Bering Seas were dry land, exposing a large land mass that stretched from the Verkhoyansk Mountains of eastern Siberia through present-day Alaska and east to the MacKenzie River in northwest Canada. During even the coldest part of the Last Glacial Maximum, this area, known as Beringia, remained snow and ice free. It was an extension of the Mammoth Steppe that extended across Eurasia from the Iberian Peninsula to the Wisconsin Glacier in northern Canada. In spite of the extreme cold and harshness of the conditions, Beringia remained habitable to mammalian life and became a northern refugia for megafauna and early humans, as evidenced by the findings in the Bluefish Caves, Swan Point, and Broken Mammoth archaeological sites.

The Bluefish Caves are located on a limestone ridge overlooking the upper Bluefish River in the Keele Range of the northern Yukon territory, and they contain the oldest, undisturbed archaeological evidence of human habitation in Canada. The caves consist of three small cavities in which loess accumulated during the late Pleistocene, covering the bones of mammoths, horses, bison, caribou, sheep, saiga antelope, bear, lion, many other mammals, birds, and fish. Many of the bones of the larger animals exhibit butchering marks and evidence of percussion and flaking by stone tools.

Also present in the loess are the lithic fragments of flint microblades and chisels that have been radiocarbon dated to between twenty-five and twelve thousand years ago. The artifacts and remains indicate that the Bluefish Caves were likely sporadically used by small hunting groups for many millennia.

A bit further west, at the Swan Point archaeological site in the Tanana River valley in east-central Alaska, there is evidence of human occupation in the form of bifacial stone and ivory tools dated to approximately fourteen thousand years ago. At the Broken Mammoth archaeological site, also in the Tanana River valley, charcoal from hearths has been dated to over twelve thousand years ago. This evidence suggests that early humans were in Alaska and the Yukon territory by at least fourteen thousand years ago and possibly as early as twenty-five thousand years ago.

Migration of the people of Beringia into the rest of North America, however, was hampered by the presence of two massive ice sheets, known as the Laurentide and the Cordilleran, that blanketed much of Canada and the northern United States during the Last Glacial Maximum.

The Laurentide ice sheet extended from the MacKenzie River and the Rocky Mountains in Canada, across the Canadian Arctic Archipelago to Greenland, and reached as far south as Pennsylvania and Cape Cod, while the Cordilleran extended from the Aleutians through southern Alaska along the western slopes of the Rocky Mountains and Pacific Coast as far south as Washington state. Even though sea levels were more than one hundred metres lower than today, the Cordilleran ice sheets extended right to the ocean's edge in most areas, covering much of the coastal areas.

Separating the two ice sheets was the American Cordillera, a chain of mountain ranges comprising an almost continuous sequence of mountains that stretched from Alaska to the southern tip of South America, sometimes referred to as the Continental Divide. Like the Eurasian Alpide Belt, they are orogenic, being formed by the uplifting of the North and South American continental plates as the Pacific plates slide beneath them. They are still being formed and are seismically active today, forming the eastern section of the Pacific Rim of Fire.

During the Last Glacial Maximum, the combination of the American Cordillera and the two massive ice sheets presented a virtually impenetrable barrier, preventing the migration of Modern Humans out of Beringia into the southern reaches of the Americas. But as the earth's climate began to moderate, ice-free corridors began to appear, one in the interior, along the eastern flank of the Rocky Mountains, known as the Western Ice-Free Corridor, and the other, a sea route, along the Pacific Northwest coast. While the sea route was the first to open, both routes ultimately provided early humans with access to the more southerly reaches of the Americas for the first time.

Until very recently, most archaeologists supported the theory that the first immigrants to the Americas were megafauna hunters from the Mammoth Steppe in Alaska that followed the big game down the Western Ice-Free Corridor into the North American interior. Archaeological remains of these early settlers, known as the Clovis People, are widely distributed throughout the south-central United States, and for many years, Clovis sites yielded the oldest evidence of human occupation in the Americas, dating to approximately thirteen thousand years ago.

The Clovis First theory, however, is somewhat problematic as the Western Ice-Free Corridor did not open until about thirteen thousand years ago, and some argue that it may have been another millennia or two before enough flora and fauna had become re-established in the corridor to support migrating herbivores and the human population that preyed upon them.

Furthermore, archeological sites along the west coasts of North and South America have yielded human artifacts that predate the Clovis sites by one to two millennia, persuading most archaeologists to consider the possibility that the first human settlers in the Americas came via the Pacific Coastal route.

Beginning about eighteen thousand years ago, warming conditions began to expose stretches of coastline along southeast Alaska and northwest British Columbia, which would have provided access to stretches of coastline from the sea. But even when the ice began to retreat from the coastlines of southern Alaska and the northwest coast of British Columbia, it would not have exposed a coastal plain that could be easily traversed on foot. Much of these coastlines consist of rocky points and bluffs that can extend for several kilometres, and such areas would have been difficult to traverse on foot. So, when the first peoples began to migrate down the northwest coast into the Americas, they would have had to, at least occasionally, travel by boat. Furthermore, their boats would have had to be fairly large and seaworthy as the North Pacific is a challenging sea to navigate even today, particularly along the coastline where two- to three-metre waves and swells continuously crash against the predominantly rocky shores. These early immigrants must have been seasoned sailors to navigate along stretches of the Pacific Northwest coastline as they are some of the most treacherous on earth.

It would appear, therefore, that the first settlers of the Americas were seafarers that came by boat along what has been dubbed by archaeologists as the Kelp Highway, so named because of the large beds of sea kelp that can be found along the entire Pacific Northwest coastline.

The *Kelp Highway Hypothesis* asserts that people occupying the coast of eastern Siberia migrated, mostly by boat, northeast along the Siberian coast and the Aleutian Islands until they reached the south shore of Beringia. From there, they proceeded along the coast of Alaska and British Columbia and down the west coast of the United States, Central America, and South America, slowly populating the west coast of the Americas as they went.

As they travelled, these first immigrants encountered rugged, inhospitable shorelines where mountains dropped precipitously into the sea, and remnants of the Cordillera ice sheet still lingered at the ocean's edge. Occasionally, however, they would find more hospitable, ice-free bays and inlets where they would have settled and remained for a time—for years, possibly even centuries—until their numbers grew, and local resources were exhausted. Then, they would continue on. As the migration continued, some stayed behind and continued to occupy the settlements for millennia; others moved ever southward along the coast.

The Kelp Highway hypothesis remains conjecture, but there is mounting archaeological evidence in support of the theory. Much of this support comes from the discovery of a signature stone point, known as the *Western Stemmed point*, which was found at several sites around the Pacific Rim, all dated to approximately the same time period. The Western Stemmed point is characterized by a stem or projection from the base of the point used for hafting the point onto an arrow or spear shaft.

The Western Stemmed point tradition was practiced by people along the western Pacific Rim, including the incipient Jomon culture of Sakhalin Island, in the Sea of Okhotsk, where a Western Stemmed point has been dated to sixteen thousand years ago, and another from the Ushki archaeological site on the Kamchatka Peninsula in Siberia, dated to approximately fourteen thousand years ago.

A Western Stemmed point.

In the Americas, stone tools and points made in the Western Stemmed point tradition have been found at the Lind Coulee and Cooper's Ferry archaeological sites in Washington state, the Channel Islands off the coast of California, and as far south as the Amazon lowlands in South America.

Some of the oldest evidence of early immigrants following the Kelp Highway are the submerged stone rectangular structures found around Haida Gwaii, an archipelago along the Pacific Northwest coast of British Columbia. One of the structures, located off the east coast of Moresby Island in Juan Perez Sound, appears to have been a weir for trapping salmon. The weir consists of a long, submerged stone wall that was constructed perpendicular to a stream, thereby creating an artificial channel through which migrating salmon would have passed as they swam towards the stream during spawning season. The weir is at a depth of 122 metres today but would have been at about sea level when it was constructed. Other rectangular rock formations found on the seabed nearby appear to be the foundations of dwellings, leading archaeologists to speculate that the site may have been a seasonal, salmon-harvesting site right on the seashore at the estuary of the stream. The weir is estimated to have been constructed almost fourteen thousand years ago.

Further south of Haida Gwaii, at Triquet Island on British Columbia's central coast, lies the site of an ancient Heiltsuk village that has evidence of more or less continuous occupation from about fourteen thousand years ago to fairly recent times. In fact, the Heiltsuk people lived at the site until the arrival of European fur traders at the coast and the establishment of a trading post by the Hudson's Bay Company during the nineteenth century, whereupon they moved their village to nearby Bella Bella. The modern descendants of the Heiltsuk people have an oral tradition about the village site that talks of how long ago their ancient ancestors settled on Triquet Island because it remained ice free while ice and snow blanketed the inland areas.

The excavation on Triquet Island has revealed much about the people living at the village site and how their lifestyle changed over time. For the first seven thousand years or so, the people hunted and ate large mammals, especially marine mammals like seals and sea lions. But about 6,700 years ago and, again, 5,600 years ago, it appears the village was struck by tsunamis. Evidence of habitation drops off after the earliest tsunami, which could mean the site went unused for a time, only to be repopulated a millennium or so later by people with different dietary preferences. Around the time of the second inundation, the villagers diet seems to have suddenly shifted to seafood as there is evidence of fish and shellfish processing starting around that time. The beach nearby the village has been altered with fish traps and stone-walled clam gardens and a five-metre-deep midden runs for seventy metres between the beach and the village site, representing some seven thousand years of discarded trash and artifact accumulation.

The village site has produced some unique artifacts, including a wooden spear or harpoon launching atlatl, compound fish hooks, and a six-thousand-year-old wooden hand drill for lighting fires. A cache of stone tools was found close to one of the hearths, conjuring an image of the people sitting around the campfire as their salmon steaks sizzled on the hearth stones, knapping their stone tools, and engaging in storytelling and conversation.

After a time, as the migrants continued to move southward, they reached southern British Columbia and the northwest United States, where they came to the southern limit of the Cordilleran ice sheet. They encountered large estuaries at the mouths of great rivers like the Fraser, Columbia, Sacramento, and Colorado, and some settlers followed these rivers inland to settle in the interior regions of the western United States.

On the north shore of the Olympic Peninsula in Washington state, a bone spear point was found embedded in a rib bone of a mastodon at the Manis archaeological site, dated to 13.8 thousand years ago. The spear must have been thrown or thrust with considerable force as it penetrated the bone by more than two centimetres after passing through a thick hide and as much as thirty centimetres of muscle tissue.

In the interior of Washington state, on the Columbia Plateau, there are several archaeological sites, such as Lind Coulee, Marmes Rockshelter, and Cooper's Ferry, all yielding evidence of human habitation dating to more than eleven thousand years ago. All of these sites are accessible via the Columbia River system, and archaeologists are convinced that the inhabitants would have arrived in the region from the west coast. At the time, the southern edge of the Cordilleran icefield lay in about the middle of Washington state, so the Columbia River would have been the first of the major ice-free west coast river systems encountered by the migrants, affording access into the interior of the continent.

The Lind Coulee archaeological site, located near Warden in the interior of Washington state, dates to approximately eleven thousand years ago. Amongst the artifacts recovered at the site are several bone tools, needles, and points, as well as stone points knapped in the Western Stemmed point tradition. Other stone-cutting and scraping tools illustrate a sophisticated proficiency in working stone. Excavations also uncovered stone palettes that had been used for grinding red ochre.

The Marmes Rockshelter is an archaeological site near the confluence of the Snake and Palouse Rivers in southeastern Washington state. Evidence from the Marmes site indicates that the occupants harvested mussels from the nearby Snake River and used the atlatl to hunt elk, deer, and beaver. The excavation also unearthed graves, which included spear points and beads carved from shells. The majority of the shells had holes drilled through them, indicating that they had been used to make necklaces. The stone projectile points found at the site were of the same Western Stemmed tradition and were made of chalcedony, chert, and even agate, which would have come from the seashore. The finds included mortars and pestles for grinding grains and seeds and stone scrapers for use in tanning hides.

The Cooper's Ferry archaeological site, referred to as the ancient village of Nipéhe by the local Nez Perce, is located within a terrace at the confluence of Rock Creek and the lower Salmon River of western Idaho. Remarkably, in August 2019, anthropologist Loren Davis and a team from Oregon State University published findings from the site that date human occupation to as early as sixteen thousand years ago.

The Cooper's Ferry site has yielded almost two hundred stone artifacts, including projectile points, flake tools, and bone fragments from large mammals, along with evidence of a fire hearth, a food processing station, and other domestic activity stations. Many of these artifacts have been dated within the range of sixteen to fifteen thousand years ago, placing Cooper's Ferry amongst the oldest human habitation sites found to date in the Americas.

Another west coast migrant settlement was discovered at the Paisley Cave complex in Oregon. The complex is a system of four caves in the arid, desolate Summer Lake Basin of south-central Oregon. Fossils and artifacts were found at the same stratigraphic level as a small rock-lined hearth some two metres below the modern surface, along with a large number of bones from waterfowl, fish, and large mammals, including an extinct camel and horse. Human remains found at the caves have been dated to between 12.75 and 14.29 thousand years old, and an analysis of the DNA obtained from human coprolites found at the site revealed they were of east Asian descent.

Human remains, dated to thirteen thousand years ago, have also been found at a site on Santa Rosa, one of the Channel Islands off the coast of California. The projectile points found at Santa Rosa are of the Western Stemmed tradition, and the mere presence of people on the Channel Islands nearly thirteen thousand years ago provides further credence to the hypothesis that these first immigrants were proficient seafarers.

Monte Verde is an archaeological site near the city of Puerto Montt in southern Chile, located on the banks of Chinchihuapi Creek, about sixty kilometres from the Pacific Ocean. In spite of its location, some thirteen thousand kilometres from Beringia, Monte Verde is one of the oldest, reliably dated, human settlements in the Americas, possibly occupied as early as 14.8 thousand years ago. It is a remarkable site from an archaeological perspective because of its exceptional preservation. The site was submerged in an anaerobic bog environment shortly after being occupied, which inhibited the bacterial decay of organic material and preserved many perishable artifacts and other items throughout the millennia.

It is estimated that the Monte Verde site was occupied by about twenty to thirty people that lived together in a small hamlet, consisting of a main lodge and a few smaller huts. The lodge was a six-metre-long, tent-like structure built alongside a creek and constructed with a wooden pole frame covered with animal hides. The interior of the lodge was partitioned into separate living quarters with hides suspended from the framing poles, and each of the living quarters had a brazier pit lined with clay. The foundations and floors of the small huts were of a size to suggest they may have been for specific tasks, such as the preparation and administering of medicines and dry storage for food and other supplies. Near the lodge and huts were two large community hearths, around which were quantities of stone tools and remnants of spilled seeds, nuts, and berries, leading one to speculate that they may have been used for socializing, celebration, or providing light and heat for fabricating tools and weapons. Nearby was a preserved chunk of meat, which DNA analysis revealed, was from an extinct elephant-like animal known as a *Gomphothere*. A nearby midden indicates that the people enjoyed a diet of llama, gomphothere, and shellfish plus a good variety of vegetables, seeds, berries, and nuts.

The remains of forty-five different edible plant species were found within the site, and researchers who analyzed the residues of the hearths reported nine different species of seaweed and marine algae, some of which are known to have been used by prehistoric people to make medicines and healing compounds. The seaweed samples were dated between 14.22 and 13.98 thousand years ago.

Continuing south, archaeological evidence indicates that the Modern Human migrants reached the southern tip of South America by about twelve thousand years ago.

Tierra del Fuego is an archipelago off the southernmost tip of Patagonia, separated from the South American mainland by the Straits of Magellan. The archipelago was named by Ferdinand Magellan, the first European to explore the area in 1520, and he chose the name, which means Land of Fire. This seemed an appropriate name because of the sightings of what he believed were bonfires lit by the indigenous people living along the strait.

When Charles Darwin passed through the strait about three centuries later, he socialized with the natives on several occasions. In his first book, *Voyage of the H.M.S. Beagle*, Darwin wrote,

"We had an interview at Cape Gregory with the famous so-called gigantic Patagonians, who gave us a cordial reception. Their height appears greater than it really is, from their large guanaco mantles, their long flowing hair, and general figure: on an average, their height is about six feet, with some men taller and only a few shorter; and the women are also tall; altogether they are certainly the tallest race which we anywhere saw. In features they strikingly resemble the more northern Indians, but they have a wilder and more formidable appearance: their faces were much painted with red and black, and one man was ringed and dotted with white paint."

These early Patagonians appeared well-adapted to the harsh climate as many were naked even during near freezing temperatures. Darwin reported observing a mother who was nursing her naked infant as freezing rain streamed down the infant's body.

Across the strait from Tierra del Fuego is a large volcanic plateau called Pali Aike, formed by lava flows from the eruptions of volcanoes located in the southern Andes mountains to the west. Fell's Rockshelter and the Pali Aike Caves are two archaeological sites located on the plateau, and both show evidence of occupation by humans about twelve thousand years ago. An analysis of the dental patterns of human skulls found at Pali Aike showed them to be consistent with a northeast Asian ancestry, and among the stone tools found were stone bolas and fishtail spearpoints, a form of stone point that resembles the Western Stemmed points with pronounced shoulders above a clearly shaped stem.

The Pali Aike caves mark the most southerly point reached by the ancient people that left Africa about sixty-five thousand years ago and began their wandering migrations along the coastline of south Asia. Over the course of their migrations, they traversed over sixty thousand kilometres of coastline and settled in all the major land masses of the world, save Antarctica.

Into the Interior of the Americas

As the coastal migrants were making their way down the west coast of the Americas, the American Cordillera afforded little or no access into the central regions of the continents. Even today, there are only a few passes through these mountains, and fourteen thousand years ago, these would have been blocked by glaciers. However, this changed when the coastal migrants reached the Isthmus of Panama where they were able to cross to the Atlantic side of the continent, setting foot on the shores of the Caribbean Sea for the first time. And from there, they dispersed north and south around the coastline into the eastern regions of North and South America.

Those that chose the south coast, settled along the shores of Columbia, Venezuela, the Guianas, and northeast Brazil, reaching the estuary of the Amazon River about twelve thousand years ago.

The Caverna da Pedra Pintada is an ancient archaeological cave site that is located in the rain forest along the Amazon River near the town of Monte Alegre, not far from the river's estuary. The lowest levels of the cave, dated to approximately eleven thousand years ago, yielded charred floral and faunal remains, stone tools, and spear points, revealing that the earliest visitors had adapted to a hunter-gatherer lifestyle in the tropical rain forest and lived on a diet of fruit, palm seeds, and Brazil nuts as well as fish, shellfish, birds, and reptiles. The Pedra Pintada cave also contained several wall paintings, believed to be amongst the earliest cave art discovered in the Americas.

Archaeological evidence from cave sites in southeast Brazil and Argentina, suggests that early humans dispersed throughout South America by following the same pattern they had followed since leaving Africa, migrating along the coastlines and then entering the interior regions along river valleys or habitable forested regions. By eight to nine thousand years ago, hunter-gatherer societies occupied most of the habitable regions of the South American continent.

The First Floridians

While people were dispersing into the South American continent from the Caribbean shores, others were migrating northward, settling along the eastern coastlines of Central America and the southeastern United States.

On the northeast coast of the Yucatan Peninsula, there is a vast network of underwater caves known as Ejido Jacinto Pat, located about twenty kilometres north of the City of Tulum. Around the time of the Last Glacial Maximum, the caves were mostly above sea level and visited by giant ground sloths, elephant-like gomphotheres, saber-toothed cats, and several other animals, large and small. In one of the larger caverns, known as Hoyo Negro, the complete, well-preserved skeleton of a young girl was found lying on the floor of the cave, surrounded by the remains of a mastodon. Both have been dated to about fourteen thousand years old, indicating that the coastal migrants had reached the Yucatan by that time.

Further north, along the shores of the Gulf of Mexico and about 250 kilometres inland and north of the city of Austin, Texas, the Buttermilk Creek archaeological site has yielded evidence of human occupation dating back over thirteen thousand years. The site is an open-air encampment along the Buttermilk River, and it is just one of many in the area that show occupation by early human hunter-gatherers for several millennia. The Buttermilk Creek Complex, along with other nearby sites in Texas, is believed to be the area where the Clovis culture evolved from an older tool industry, most likely the Western Stemmed tradition, that predated Clovis by as much as twenty-five hundred years.

Further along the Gulf Coast is the state of Florida, where numerous archaeological sites have been found that provide insight into the late Pleistocene environment of Florida and the lifestyle of the people that settled there some twelve to fourteen thousand years ago.

Florida's environment at the end of the Pleistocene was very different from what it is today. Its climate was much cooler and drier and, with sea levels at least one hundred metres lower, the Florida peninsula had about twice the land area. The landscape was mostly savanna with a few scattered lakes and rivers, and during this time, the Florida peninsula was home to the greatest diversity of terrestrial vertebrates the North American continent has ever held. Herds of American mastodon and mammoth grazed on the Florida steppe foliage along with other herbivorous megafauna like the giant ground sloth, the huge and heavily armoured armadillo, giant tortoise, antelope, bison, camel, deer, and horse. On the peninsula, fresh water was available only at a few scattered lakes and rivers, and animals congregated at these water holes to drink, providing human hunters with ample opportunities to hunt them. Humans, however, didn't have the huge herbivores all to themselves. Other carnivores, like the enormous short-faced bear, the sabre-toothed cat, wolves, and the American lion, one of the largest cats to have ever existed, all preyed upon the same herbivores, and most certainly upon any human that happened to stray from their group. In addition to the megafauna, the remains of snakes, raccoons, opossums, giant two-hundred-kilogram tapirs, muskrat, turtles, and both fish and shellfish have been found at the many archaeological sites throughout the state.

Florida was one of the last holdouts of Pleistocene megafauna in the Americas, and the first humans that arrived there appear to have been quick to adapt to the new environment and capitalize on the big-game hunting opportunities it presented. The people thrived in this new environment, and their numbers increased quickly as archaeological sites abound throughout the Florida peninsula, dating to between fifteen and twelve thousand years ago. In addition to the numerous land sites, human-related artifacts have been found in flooded river valleys as much as five metres under the Gulf of Mexico, and suspected sites have been identified up to thirty-two kilometres offshore under twelve metres of water.

Thousands of stone tools and points have been found at numerous sites throughout the Florida peninsula that show an evolving tool industry that lasted from fourteen to eight thousand years ago, the time known as the *Paleo-Indian period*. One of the more distinctive points from the latter part of this period is the *Bolen point*, a small to medium-sized stemmed and notched point. While the Bolen points show a marked refinement and are perhaps more aesthetically appealing than the Western Stemmed points, their resemblance is undeniable.

A Bolen projectile point from Florida.

The Florida sites have yielded a treasure trove of artifacts made from stone, ivory, bone, antler, shell, and wood, including eyed needles, double-pointed pins, and part of a mortar and a throwing stick that are both carved from oakwood. One of the more intriguing finds was a mammoth or mastodon ivory foreshaft, which allowed for the quick connection of a projectile point onto a wooden shaft. One end of the foreshaft was permanently attached to a projectile point with pitch and sinew, and the other end was pointed for pressure-fitting into a wooden shaft. Such a design meant that once the point had entered the animal, the shaft could be removed, a new point quickly attached, and the weapon was ready for reuse.

The Clovis People

Between thirteen and fourteen thousand years ago, as the early Paleo-Indian people in the southeast of the United States began to disperse into the heartland of the country, the Laurentide ice sheet was beginning to retreat, opening the Western Corridor along the eastern edge of the Rocky Mountains. The southern edge of the ice sheet, which had extended into most of the northern states, retreated northward into Canada, leaving behind massive depressions in the earth's crust, some of which filled with water from the melting glaciers to form the Great Lakes and the lakes of central Manitoba.

As the Western Corridor began to open and the Alaskan steppe ecosystem expanded south along the British Columbia–Alberta border, the Beringia megafauna and human bands began to migrate southward. By about thirteen thousand years ago, herds of horse and bison were grazing east of present-day Edmonton, Alberta, and within a few centuries, bands of migrants from Beringia were hunting herds of bison in southern Alberta and on the great plains of the American west.

As these northerners crossed into the northern states of Idaho and Montana, they began to quickly disperse into the American heartland. Some moved east along the southern fringes of the retreating icefields to hunt the woolly mammoth. There are many archaeological *kill sites* that contain the remains of butchered mammoth throughout the northern American states of Montana, North Dakota, Minnesota, Wisconsin, Michigan, and Pennsylvania, dating to thirteen thousand years ago. Such kill sites include the Mud Lake, Schaefer, and Hebior sites in Wisconsin that contain the remains of butchered mammoth and quantities of stone butchering tools that have been dated to this time. To the southeast, along the Ohio River valley in southwest Pennsylvania, the Meadowcroft Rockshelter shows evidence of use by early Paleo-Indian bands for butchering animals from possibly as early as sixteen thousand years ago, although the majority of archaeologists dispute such an early date. Nevertheless, if the early dating of the Meadowcroft artifacts is indeed accurate, the site predates almost every other archaeological site in the Americas by one to two thousand years and is the strongest evidence yet found that supports the Solutrean hypothesis.

The Meadowcroft site lies at the southern edge of the Laurentide ice sheet when it was at its maximum extent at the end of the Last Glacial Maximum. Many, if not most, of these sites show no evidence of prolonged habitation, suggesting that the region was populated with bands of hunting and gathering people that would make a major kill, like a mammoth, and butcher it where it fell. After spending enough time at the site to consume the animal and process the hides into clothing, footwear, or shelter covers, they would take what they could carry and continue wandering in search of their next prize. These roving bands left little trace of their presence other than the remains of their kills and a few abandoned tools. They had no time or inclination for art or carvings and typically left the massive ivory mammoth tusks behind with the bones.

By about thirteen thousand years ago, small bands of hunter-gatherer Paleo-Indian people lived throughout the heartland of the United States, having come from dispersals out of the southeastern states and from the north via the Western Corridor. They lived a subsistence lifestyle, and their overall population was relatively small. Bands were widely scattered throughout the continent, and they moved frequently, following herds of game during their migrations or moving when their preferred food sources became depleted. Bands often moved every few days, possibly travelling three hundred kilometres or so in a hunting season. Early Paleo-Indian bands engaged in this nomadic, hunter-gatherer lifestyle from early spring until fall, but as the weather cooled and snow began to fall, they would break into smaller groups and retreat to more sheltered sites to wait out the winter. If the winter was unusually long and harsh, or if food supplies proved inadequate, some would perish in their lodges before spring.

It was at this time that the Clovis culture emerged amongst the Paleo-Indian people of the American heartland. The culture was named after the Blackwater Draw site near Clovis, New Mexico, where the first distinctive stone and ivory tools, including the beautifully crafted *Clovis points,* were first found.

The Clovis point has become the iconic symbol of the American Paleo-Indian and is considered by some to be the finest example of craftsmanship from the Palaeolithic period. The Clovis point is bifacial and pressure-flaked, typically made from chert, jasper, quartzite, chalcedony, or obsidian. When these materials are struck, small waves that resemble a ripple in a pond are left in the stone at the point of impact, making an extremely sharp edge. Apart from being easy stones to work with, the material used to make the points was often strikingly beautiful and came in many different colours. Clovis points can range from 5 to 25 centimetres long and are typically 2.5 to 5 centimetres wide, but their most distinguishing feature is the flute, a small groove roughly 2 centimetres long at the base, which facilitates fastening the point onto a shaft.

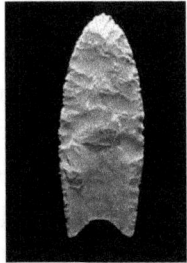

The iconic Clovis projectile point.

Just as the appearance of the Aurignacian culture marked the beginning of a change in the way anthropologists refer to groups of ancient people throughout Eurasia, the appearance of the Clovis culture changed the way groups of indigenous people of the Americas are referenced. Prior to the appearance of the culture, all indigenous people in the Americas were referred to as Paleo-Indian, but following it, specific group names were introduced. Hence, the term Clovis refers to a technological tradition, a culture, and the people who practiced it. The Clovis people were the first indigenous group of the Americas to be so recognized, and the custom has continued into modern times as groups of indigenous peoples throughout the Americas—the Sioux, the Mohawk, the Cheyenne, and so on—all have names that identify their unique cultural heritage.

The Clovis point was manufactured in various sizes depending on the type of game being hunted. Some were as much as twenty-five centimetres in length, designed to kill the large megafauna that roamed the Americas during the Clovis period. At the time, there was a large population of exceptionally large animals that included the giant sloth, the short-faced bear, tapirs, peccaries, the American lion, giant tortoises, the American cheetah, sabre-toothed cats, dire wolves, antelopes, camels, llamas, giant bison, the stag moose, the shrub ox, musk ox, pronghorns, horses, mastodon, armadillos, the giant beaver, giant condors, and various species of the huge raptor *Teratornis*. Along the Pacific Northwest coast of British Columbia, a three-metre-long sabre-toothed salmon could be found swimming up the rivers to spawn in the fall.

The Clovis culture was brief, lasting only from about 13.2 to 12.8 thousand years ago, but it spread rapidly throughout most of the United States and southward as far as Colombia and Venezuela. The rapid spread of the Clovis culture was not due to the migration and settlement of Clovis people, but rather through trade and cultural exchanges that took place whenever wandering hunter-gatherer bands encountered one another. The reason bands would abandon their traditional toolmaking industry to adopt the Clovis tradition appears to be purely aesthetic, as the Clovis points and blades offered no particular functional advantage over earlier toolmaking traditions, but they were truly works of art. It is believed that Clovis points, in particular, were highly prized by Paleo-Indian hunters, as a finely crafted Clovis point would have been a source of pride, earning the admiration and respect of fellow hunters and band members. Hence, the rapid spread of the Clovis culture and the universal adoption of the Clovis industry in so short a time period.

Following the Last Glacial Maximum, the earth's climate began to moderate, and the Laurentide and Cordilleran ice sheets started to melt and retreat. Vegetation in the northern hemisphere responded to the warming climate with the expansion of forests and grasslands. With the melting of the ice, fresh water began to flow into the oceans, and sea levels began to rise again. But the warming trend in the northern hemisphere appears to have suddenly reversed, and temperatures fell abruptly. Within just a few decades, the earth's average temperature dropped by as much as 6°C in some places in Europe and North America. Glacial melting ceased, and the ice and glaciers began to advance once again.

This sudden change in climate is known as the *Younger Dryas event*, which lasted from approximately 12.9 to 11.7 thousand years ago. It is the youngest of three rather sudden stadials that occurred near the end of the Pleistocene and is named after the Arctic wild flower, *Dryas octopetala*, an indicator plant that thrived in the sudden cold environment created by the event.

The prevailing theory is that the Younger Dryas was caused by significant reduction or shutdown of the North Atlantic conveyor, a strong ocean current that circulates warm tropical water from the Gulf of Mexico northward along the eastern coastlines of the United States and Canada. In the North Atlantic, the massive current splits into two streams, with the northern stream crossing to northern Europe and the southern stream recirculating off west Africa. The current, now referred to as the Gulf Stream, warms the climate of the east coast of North America from Florida to Newfoundland, Iceland, and the west coast of Europe.

Prior to the onset of the Younger Dryas stadial, the warming of the earth's climate and melting of the icefields and glaciers in Canada and northern Europe released vast amounts of fresh water, which flowed down to the coasts and spread out across the North Atlantic Ocean. As a thick layer of fresh water formed over the ocean's surface, it had the effect of pushing the North Atlantic conveyor southward, possibly interrupting the flow of its northern branch entirely. This shifting would have occurred suddenly, and a sharp cooling of the northern climate over just a few decades would be quite consistent with such a shift in ocean currents. With the drop in temperatures, glacials would begin advancing once again, and the runoff of fresh water would cease. Over time, only a millennium it seems, the freshwater layer over the North Atlantic was reabsorbed and the conveyor was reestablished, following which the climate in Canada and northern Europe warmed once again.

The Younger Dryas event is believed to have triggered the sudden disappearance of most of North America's large animal species that occurred between thirteen and twelve thousand years ago. At most archaeological sites throughout North America, there is a stratigraphic layer of organic-rich earth, referred to as the *black mat*, that was laid down during the Younger Dryas event, indicating a massive die-off of the vegetation that had expanded northward in response to the previous warming period. This layer or mat covers the surface on which the last remnants of the Pleistocene megafauna are recorded, and many scientists believe that it was the sudden loss of life-sustaining vegetation that was the primary factor that led to the extinction of the herbivorous megafauna and the predators that hunted them.

Historically, much of the blame for the extinction of the Pleistocene megafauna has been assigned to overhunting by the early Paleo-Indians, in particular, the Clovis people, but the stratigraphic record indicates otherwise. Or perhaps, climate change and overhunting were both factors.

Not all the North American mammalian fauna disappeared during the Younger Dryas event, however. The ancestors of today's reindeer, musk ox, caribou, and other Arctic animals survived the event and migrated north as the ice sheets finally retreated for the last time. Another species that survived was *Bison antiquus*, ancestor of the modern buffalo. This ancestral bison was about 2.3 metres tall and weighed about sixteen hundred kilograms, making it about 20 percent larger than today's buffalo. Tip to tip, its horns measured about one metre, a nasty fellow to deal with at close range. The numbers of these Pleistocene bison appear to have grown substantially following the Last Glacial Maximum until about ten thousand years ago when they began to disappear, only to be replaced by the modern, somewhat smaller, buffalo that were better adapted to life on the prairie grasslands that were appearing on the Western plains. The buffalo thrived, increasing in numbers rather quickly, and by five thousand years ago, the vast buffalo herds of legend had appeared on the great plains of the American West.

It also appears that the Clovis culture, itself, ended rather abruptly during the Younger Dryas event as no Clovis artifacts have been found in the layers above the black mat. The preferred prey of the Clovis people was the woolly mammoth and mastodon, which completely disappeared during the event, and perhaps, the Clovis people had become too specialized in their hunting practices to adapt quickly enough to the more limited numbers and types of game that remained.

Oceanographers today warn that Dryas stadial cycling is known to have occurred at least three times throughout the Pleistocene and could occur again in modern times. With the recent increase in the rate at which the polar ice caps and mountain glaciers are melting, a layer of fresh water is, indeed, building over the North Atlantic, and there is a reasonable probability that a southward shift in the Gulf Stream could occur within a few decades. It is rather paradoxical that at a time when scientists are sounding the alarm about the rapid warming of our planet and the associated climate changes, one of the major effects of global warming may in fact be the triggering of a fourth Dryas event and the shifting of the Gulf Stream southward, which would plunge most of Europe and northeast America into a deep freeze.

The Folsom People

Just as the Clovis culture was coming to an end, a new culture emerged, known as the Folsom complex. And whereas the Clovis people had developed weapons and hunting strategies that centred around the woolly mammoth and mastodon, the Folsom people focused on the pursuit of the herds of *Bison antiquus* that were beginning to appear in large numbers throughout the central United States.

The Folsom people were nomadic hunter-gatherers and lived in small, highly mobile family groups, travelling long distances across the plains as they followed the migrating bison. They were skilled hunters, but the bow and arrow were not part of their weaponry. Folsom hunters had only close-range weapons with which to kill their prey. Based on the number of bison represented at their kill sites, it is likely that the Folsom hunters used natural terrain features such as gullies or steep-sided valleys to confine small herds of the animals, allowing them to get close enough for their weapons to be effective. When they did, they typically slaughtered several of the animals before the others escaped. After such a kill, the hunters would set up camp and live for a period of time while they processed and consumed their kills.

The Folsom type site is a bison kill site along the Wild Horse Arroyo stream near the town of the Folsom, New Mexico. Evidence found at Folsom indicates that twenty-three bison were trapped and killed at the site, and radiocarbon dating on the bones indicated an average age of

12,500 years. Other archaeological discoveries also confirm that the Folsom people were formidable bison slayers as several kill sites have been found with the remains of up to fifty bison that were slaughtered and rendered during a single hunt.

The Folsom lithic tradition was characterized by the Folsom point, a distinctive projectile point that has been found, along with similarly fashioned slaughtering and butchering tools, at the bison kill sites. The Folsom projectile point was an adaptation of the Clovis point for the specific purpose of hunting bison. Although *Bison antiquus* was much larger than the modern buffalo, weighing as much as nine hundred kilograms, it was rather diminutive in comparison with a woolly mammoth, which could weigh as much as eight thousand kilograms. The bison were also much faster runners than the mammoth and travelled in large herds, whereas the mammoth moved in smaller, family groups.

The Folsom point was smaller, lighter, and a different shape than the Clovis, and in combination with the Folsom hunter's innovative method for hafting the points to a spear, it proved deadly at killing *Bison antiquus*. Folsom points were permanently fastened to a foreshaft, which, in turn, was pressure-fitted into the body of a shaft or lance. The composite spear would then be thrust or thrown into the body of an animal, and the foreshaft would either be twisted off or it would break off, leaving the point and foreshaft in the animal. As the animal struggled, the point worked its way deeper, damaging internal organs, and speeding the kill. Over time, the Folsom hunters began using the atlatl, a hand-held throwing stick, to increase the penetrating power of their spears.

The key to the lethality of the Folsom point was its wafer-thin design, which allowed it to slip between the ribs of the bison to reach internal organs. The thinness was achieved at the very end of the production process when a wide central groove, or *flute*, was created by knapping away material down the centre of the point on at least one side, and often both sides. This final step was a tricky one and would often result in breaking the point, after probably an hour or two of work to get it to that stage. But if this final step was successful, the result was a very thin and razor-sharp projectile point.

Very few permanent or semi-permanent Folsom settlement sites have been found, indicating that entire families wandered with the hunters, carrying their worldly possessions with them. When a kill was made, they would set up a temporary encampment to butcher and process their kills. Such kill sites have been found throughout the U.S. Midwest, from Texas and New Mexico in the south, through Colorado and Oklahoma, to North Dakota, Montana, and Wyoming in the north, and lithic tools left at kill sites have been found up to nine hundred kilometres away from the areas where they were sourced.

A few Folsom sites, however, do suggest that some Folsom tribes occupied certain campsites for an extended period of time. One such place is the Mountaineer site, located on top of Tenderfoot Mesa, near Gunnison, Colorado. More than sixty clusters of artifacts have been found at the Mountaineer site, including many Folsom tools, projectile points, and the remnants of structures, dating to between 11.5 and 7.8 thousand years ago.

In one section of the Mountaineer site, researchers found the remnant circle of rocks and postholes of a round structure, measuring about four metres across. Evidence suggests that the structure once had a teepee-like covering made of aspen poles and other plant material, capped with mud. The teepee had been built around a shallow basin with rocks piled around the edges, and it appears to have been a dwelling, complete with a hearth that contained rib bone fragments—perhaps from a bison—and a storage pit. Scattered around the interior were stone tools, projectile points, choppers, and an anvil for crushing bones. The bone fragments inside the structure have been radiocarbon dated to just over twelve thousand years ago. Outside the lodge was a second hearth, more stone tools, and several postholes, evidence of a windbreak or rack for hanging and drying food.

The Twilight of the Hunter-Gatherer

It was around the time that the Folsom culture was disappearing from the plains of the American West that new and radically different lifestyles were being adopted by people living elsewhere in the Americas and on the far side of the world.

In the northeastern United States, the people of the woodlands were domesticating wild squash, and in the Balsas River valley of south-central Mexico, others were planting carefully selected seeds of the wild teosinte plant, which before long would genetically modify into an ancestor of modern maize.

By about five thousand years ago, societies in the eastern and southern United States were cultivating winter squash, corn, and beans, and many had come to depend on these crops, known as the Three Sisters, as their principal source of food. When the first Europeans arrived in the northeastern states, the Iroquois nation traded three sister seeds and vegetables with the early settlers, which helped them survive the bitter New England winters for which they were otherwise ill-prepared.

And in faraway southern Mesopotamia, along the banks of the Tigris and Euphrates Rivers, city-states of tens of thousands of people each were flourishing. Humankind had entered the Bronze Age, and the combination of advanced agricultural practice, increased population density, and the fabrication of metal tools and weapons, was transforming human culture and lifestyle. Modern civilization had been unleashed upon the earth, and the course of human evolution would soon be forever changed.

But it was also at this time that massive herds of modern buffalo were appearing on the plains of western North America, and the people that lived there, the Blackfoot, the Sioux, the Cheyenne, the Comanche and many other tribes and nations, were adopting a lifestyle that revolved around the movements of herds of these majestic animals. The people of the plains came to rely on buffalo meat, hide, sinew, and bone to supply them with virtually everything they needed to sustain themselves, and they developed cultures that celebrated and revered the great beasts. Their buffalo-centric cultures and hunter-gatherer lifestyle continued into modern times and was flourishing still when the first Europeans arrived on the western plains in the nineteenth century.

As the time of the hunter-gatherer drew to a close, societies had evolved across the globe with a rich diversity of culture and customs. But in spite of this diversity and such vastly different life experiences, all the people of the time shared a similar fundamental spirituality. They all believed in the existence of a *universal essence* or *spirit* that permeated the natural world and resided in all living things. Perhaps such beliefs persisted and were so ubiquitous because the hunter-gatherer lifestyle kept people close to nature, and they were constantly reminded that they were an integral part of the mysterious, endless circle of life and death that was unfolding around them. Various forms of this ancient spirituality persist today amongst the indigenous people across the globe, and anthropologists refer to these ancient belief systems collectively as *Animism*.

In modern times, the universal essence has become associated with the creator of the universe and the inner spirit is referred to as the soul, a manifestation of the universal essence within all life forms. And throughout history, human societies have developed cultures centred on their relationship with the creator, each with their own unique way of acknowledging its existence. Some have worshipped it, some have feared it, and many have celebrated it with song and dance. But, since the dawn of human spirituality, all cultures have revered it and strived to communicate with it through shamanism, prayer, music, ritual, or other spiritual and religious practices.

The creator has been given many different names in both modern and ancient times. Throughout the world today, it is known in various cultures as *God, Jehovah or Elohim, Brahmana, Krishna,* or *Allah*.

In pre-Columbian America, the Iroquois people of the eastern woodlands called it *Orenda*, which means a "supernatural force" or "spiritual energy that permeated the earth," and the Algonquin tribes across southern Canada called it *Gitchi Manitou*, or the "Great Spirit."

And out on the great plains of the American West, the Lakota Sioux referred to it as *Wakan Tanka*, "The Great Mystery," "The Sacred," "The Divine."

PART TWO
Our Lofty Destiny

CHAPTER TWELVE

The Remarkable Mind of Homo Sapiens Sapiens

"If the brain was simple enough to be understood, we would be too simple to understand it."

...M. A. Minsky

WHEN *HOMO SAPIENS sapiens* first appeared in the Great Rift Valley of East Equatorial Africa, some two hundred thousand years ago, their cognitive capabilities were unlike anything the world had ever seen. Never before had the evolutionary process of natural selection produced such an overwhelmingly powerful survival strategy as the human brain. The enhanced intelligence of *Homo sapiens sapiens* proved so successful that, in virtually no time at all, evolutionarily speaking, our species became the dominant animal on Planet Earth. And in recent times, human populations have exploded in numbers that are now proving almost too large for Planet Earth to sustain.

The cognitive capabilities of the human mind are nothing short of miraculous, and they seem to go well beyond what would be expected if survival were the only objective of their creation. Our remarkable minds enable Modern Humans to not only survive but to also embrace our lives and enjoy the experience of living in ways that are unique in the animal kingdom. We have developed intricate languages that allow us to share thoughts and feelings. We have extraordinary reasoning powers that enable us to acquire a deep understanding of how the universe and the natural world operate and to create technological marvels that add immeasurable comfort and enjoyment to our lives. We express and enhance our emotions through artistic endeavour and contemplate the spiritual aspects of existence. Our cultures celebrate the beauty of the natural world and rejoice in the experience of being alive. No other species, extinct or extant, has ever come even remotely close to evolving with such capabilities.

But why would an animal evolve with the capacity to embrace the experience of living with such extraordinary compulsion, to strive to understand the magnificent workings of the universe, and to seek meaning in its own very existence? Are such avenues of thought an incidental aberration that comes as a by-product of increased cognitive powers? Or are they the predestined, ultimate purpose of life's evolutionary journey?

We humans are deeply divided by which of these beliefs each of us embraces.

The brain and the mind are not the same entity. The brain is the physical *host* of the mind, and the physical makeup of the brain—its size, organization, and complexity—dictate the workings of the mind. But it is the mind that is the seat of consciousness, the place where thoughts and feelings occur. All living things, whether animal, plant, or microorganism, possess both a brain and a mind—to a greater or lesser degree. It seems ludicrous to think of a plant or amoeba as having a brain, but to the extent that there is even the simplest form of centralized control or responsiveness in a living entity, it can be argued that there must also be a rudimentary brain with a correspondingly elementary form of mind and consciousness.

Much is known about the human *brain*—how it is organized, what parts of the brain process sensory data, what parts control various parts of the body, and how one part of the brain relates to another. In modern times, scientists in many different disciplines have produced functional maps of the human brain, and much is now known about how each of the brain's components work together to enable us to function as we do. Such research has led to many advances in the treatment of physical injuries to the brain and the development of cures for brain-crippling diseases.

Less, however, is known about the human *mind*. While much has been learned about how the body's chemistry can alter moods, emotions, and even thoughts in our conscious mind, almost nothing is known about how our minds formulate thoughts, generate feelings, solve problems, or imagine things that do not exist in reality.

In the course of our daily lives, our conscious minds are constantly flooded with a complex integration of instinctive impulses, emotions, and thoughts, and through a rather mysterious mental process, we make decisions and reach conclusions. For centuries, many scientists and philosophers have devoted their lives to understanding the nature of this process—the physical basis by which thoughts and feelings are manifest and, indeed, the nature of consciousness itself. While much progress has been made in understanding the function of most parts of the brain, the mechanisms whereby they work together in our mind to create feelings and ideas, solve problems, or imagine future constructs and outcomes, remains a mystery.

The Mystery of Cognition

"We are caught up in a paradox, one which might be called the paradox of conceptualization. The proper concepts are needed to formulate a good theory, but we need a good theory to arrive at the proper concepts."

...Abraham Kaplan

The cognitive capability of the human mind arises through a combination of several distinct, enabling functions that include memory, conceptualization, logical reasoning, emotion, intuition, and imagination, all of which are highly integrated in our consciousness and work in concert to create thoughts, ideas, and feelings. These, in turn, enable us to conduct our daily routines and enjoy the experience of living with an astounding creative capacity, perhaps the most differentiating hallmark of humankind.

Conceptualization is the fundamental enabler of cognition. It is the ability we all have to form and manipulate simple mental abstractions or concepts from the realities that our senses deliver to our brains. Conceptualization enables language, abstract thinking, the formulation of ideas and thoughts, and the acquisition and application of knowledge.

Like most of the more complex animals, humans interact with the physical world through the five senses: sight, hearing, touch, taste, and smell, which, during every waking moment, deliver a constant stream of data to our brains. The amount of raw data our senses deliver throughout the day is overwhelming, far in excess of what our brains can possibly absorb. But somehow, through the process of conceptualization, our brains sift through that data and extract the essential information in the form of a model, or concept, which it stores in our memory in such a way that we can retrieve it when needed. Every human brain, and quite likely the brains of most animal species, has this ability to some degree, yet how the brain does it is unknown.

When we survey our environment and observe various objects surrounding us, our minds immediately form abstract models and conceptual descriptors of the objects we see. For example, when I look out my window, I can see a utility pole standing near the corner of the street. When I stop looking out the window and describe what I have seen to someone else, I tell them there is a pole across the street, and they will conjure up a mental picture in their own mind of what I've just seen because we share the same concept or model in our brains of what constitutes a pole.

My mind also stores conceptual descriptors or attributes of the pole; it is wooden, grey in colour, and round with a thirty-centimetre cross-section, which I can also convey to anyone who is interested. In describing the pole with such words as wooden, grey, and round in cross-section, I am also using concepts, or *conceptual descriptors*, that trigger the same image in the minds of others. Each of us has a mental image of a wooden surface, the colour grey, and roundness. In like fashion, we store concepts of most of the objects we see in the course of our daily lives, and we have thousands of such conceptual models and descriptors stored in our mind.

Without conceptual modelling, the brain would essentially have to store entire images of objects, from a number of different viewpoints, in order for it to have enough information to be useful. While the brain can do this, this approach would require a great deal more storage and processing capacity to retrieve and manipulate the information. And if we had to store all such data for everything we see, hear, smell, touch, or taste, our brains would be overwhelmed with data in a very short time. By storing conceptual models of its surroundings, the brain can maintain a lot more useful information and can access, process, and communicate that information more quickly and efficiently.

Sophisticated language is only possible with conceptualization. Without the ability of the mind to form concepts and associate a name with those concepts, our vocabulary and language would be limited to a few sounds, like those of most other animals. Language enables humans to communicate efficiently with one another, to exchange information and ideas, to share experience and knowledge—all based on conceptual modelling. Thinking, itself, is made possible by conceptualization as we think in terms of concepts; logical reasoning is the process of examining relationships and associations amongst those concepts.

The human mind also takes conceptualization to the more sophisticated level of archetyping for remembering more complex objects. The archetype of a class of objects is a creation or abstraction of the mind, based on some sort of averaging of the shape of many objects in a class to create an ideal or perfect shape. Hence, we form mental archetypes or images of the perfectly shaped tree or the classic features that constitute the perfect human face or body.

Archetypical models are a highly efficient method of storing and communicating information about complex objects, and detailed descriptions of objects are often communicated by describing variations from the archetypal form. For example, one might refer to a young fir tree as having a perfect shape except for a bare spot on one side and a top that is rather spindly. By describing a fir tree this way, the recipient of the information immediately conjures up their image of the archetypal fir tree, which is more often than not universally shared, and then adds in the specified deviations from that image—that is, the missing branch in the middle and the lack of well-formed small branches at the top.

Human faces are most often described in terms of deviation from the archetype ideal. Law enforcement sketch artists, for example, work very much in this manner when making a sketch of a crime suspect whose face has been seen by a witness. The artist typically asks the witness a series of questions such as, "Was their face narrow or wide? Was their nose large or small? Were the eyes wide apart or narrowly set?" and so on, all aimed at determining how the face deviated from the archetypical face.

The average human mind carries hundreds, if not thousands, of archetypal forms in its memory, ranging from living organisms—like the human body, animals, and plants—to natural features or phenomena such as volcanoes, sunsets, or lightning bolts.

The archetypal form is the de facto most perfect form that an object can assume, and humans are somewhat obsessed with seeking out archetypical perfection in making choices in our daily lives. The closer an object conforms to its archetype, the more beautiful, attractive, or pleasing it seems to be, and given the choice of several samples of the same item, we are inevitably drawn to the one closest to the archetypically perfect specimen, whether we're choosing a bouquet of flowers or selecting an apple from the fruit stand in the market.

Philosophers have studied conceptualization and archetyping for centuries. In the fifth century BCE, the Greek philosopher Plato was among the first to recognize the role of archetyping in human perception. In his *Theory of Forms*, Plato argued that these mental archetypical concepts, which he called *Forms,* represent the most accurate or perfect representations underlying physical reality, and they are unchanging and eternal, transcending both time and space. Real objects, he maintained, are imperfect, decaying copies of their corresponding underlying Form.

Plato postulated that these Forms reside in a non-material sphere of existence, which he called the *Realm of Forms*, and that they are accessed and recognized by the mind when real-life objects are sensed. Hence, concepts and archetypes are the same for everyone, and it is that universality that enables the exchange of ideas and information.

Conceptualization, Mathematics, and Science

"Mathematics is the highest form of pure thought."

...Plato

Mathematics is arguably the purest form of conceptualization. While there are over a thousand languages spoken throughout the world today, the language of mathematics is understood by everyone, regardless of linguistic or cultural background. Mathematical concepts are universal and timeless in that they do not change as humankind's knowledge of the physical world evolves. Quite the contrary. Since its founding by Pythagoras in the sixth century BCE, mathematics has provided a consistent framework for the acquisition of knowledge of the natural world and continues to do so today.

All of the various branches of mathematics start with basic concepts that are extracted from observations of the physical world, and they provide methodologies for manipulating those basic conceptual entities to discover relationships amongst them, revealing new truths about the realities they represent. In this way, mathematics provides the framework for the expression of most scientific theories and is the foundation upon which most technological tools and devices are invented, designed, and manufactured. So accurately can we apply mathematics to describe the behaviour of physical entities that we tend to forget that they, and the scientific laws they encapsulate, are only abstract approximations to reality.

Consider, for example, Newton's law of universal gravitation, which formally stated, says that every body of matter in the universe attracts every other body of matter by a force that is proportional to the product of the two masses and is inversely proportional to the square of the distance between them. Newton developed his theory over three hundred years ago, and along with his three laws of motion, Newton's laws still provide the foundation of engineering design today. Newton's laws are taught in high school and university-level physics courses worldwide. They are so universal in their applicability and so accurate in describing the mechanistic behaviour of physical bodies that they have come to be treated as an inherent property of the natural world, a de facto reality as it were.

But this is not the case. Our scientific laws and theorems are not inherent properties of nature; rather, they are conceptual models of them. Nature does not behave in accordance with human constructs; humankind strives to conceive concepts and develop scientific laws that describe, as closely as possible, what nature is and does.

Mathematically, Newton's law of universal gravitation is written as:

$F = G \times (m_1 \times m_2) / r^2$

Where F is the force of gravity between two bodies of masses m_1 and m_2, G is the gravitational constant (6.673×10^{-11}), and r is the distance between the centres of the two masses.

The law of universal gravitation contains many simplifying assumptions about the reality it represents. First, the gravitational constant G is not a mysterious property of nature but is rather an empirical number selected to change the law of universal gravitation from a law of proportionality to a law of equivalence—that is, an equation.

In addition, the objects with masses m_1 and m_2, whether they are the moon or an object such as you on the earth's surface, are modelled as points, which is to say that the gravitational force, F, acts through the centre of the masses, and the direction and magnitude of the force of attraction and the distance r between the objects do not change with the orientation of the objects, relative to one another. It may seem a stretch that the moon, the earth, or you can be modelled as a point of mass, yet for the purposes of predicting the behaviour of material objects under the force of gravity, it is a very accurate conceptual representation.

In spite of all the simplifying assumptions they contain, Newton's law of universal gravitation along with his three laws of motion are regarded as being amongst the most significant scientific discoveries of all time and provide the theoretical foundation for scientific research and engineering design to this day. They accurately describe the orbit of planets around the sun and were fundamental in the development of the technology that landed Neil Armstrong and Buzz Aldrin on the moon and brought them safely back to earth in 1969.

Similarly, all the well-known scientific laws, from James Clerk Maxwell's theory of electromagnetic radiation to Albert Einstein's theory of general relativity, employ conceptual simplifications in order to make them mathematically tractable. The fact that such laws prove to be so remarkably accurate in predicting the behaviour of the real physical phenomena is testimony to the power of our brains conceptualization capabilities and, of course, mathematics!

Conceptualization, Beauty, and the Arts

"Beauty is the purest feeling of the soul. It arises when the soul is satisfied."

...Amit Ray

We are blessed to live on a planet that is filled with beautiful natural wonders. The earth hosts a vast array of diverse environments, from the parched, dry desert lands of the Sahara or the glaciered expanses of Antarctica, to the tropical rain forests of the Congo and Amazon. And there is beauty to be found in each and every one of them.

Humans, it seems, can find beauty in almost everything around us, and we seek it out as much as possible. We'll go out of our way to watch a sunset or to see a waterfall cascading over a high, rocky ledge. We blaze trails through our wild spaces for the sole purpose of experiencing the beauty that abounds in such places, stopping to admire the tiniest yellow mushroom growing out of an emerald green, moss-covered log or a shimmering snow-capped mountain that rises majestically out of a forest of crimson, orange, and yellow autumn leaves.

We find or add beauty in almost every aspect of our daily lives and strive to surround ourselves with beautiful things. Much of our culture is centred on creating and enjoying beauty in many different forms. We architect magnificent buildings and beautify our cities with green spaces, art museums, and public sculpture that serve to add enjoyment and enrichment to our daily lives. We wear beautiful clothes and adorn ourselves with objects made of precious gems and metals. We decorate our homes with artwork, sculpture, and other beautiful objects and surround our houses with gardens of flowering plants and trees.

And beauty is not limited to our visual experiences. We find beauty in the sweet, soft tones of a songbird; in the delicate fragrance of a rose; in the soft, smooth feel of an animal's fur. We embrace beauty in art, sculpture, music, dance, poetry, and prose.

We live life surrounded by beauty, and it enriches our lives enormously. The appreciation and enjoyment of beauty seems to be related to the very purpose of life itself. A life without beauty would be a very dreary existence, indeed—hardly worth living.

Yet though we all recognize beauty when we experience it, we are hard-pressed to explain what beauty actually is. What is it that makes one thing beautiful while another is not? And what is the nature of beauty that it can be recognized in such diverse things as a sculpture, a sunset, a musical concerto, the feel of a fabric, the smell of a flower, or the rhythm and words of a poem?

Throughout history, many scientists and philosophers have strived to understand the nature of beauty and why some sensory experiences are perceived as beautiful while others are less so.

In his dialogue *Hippias Major*, Plato argued that beauty is a concept, or Form, in itself. Like a mathematical theory or an archetypal shape, he proposed that beauty is an absolute truth that exists in the Realm of Forms, and it is unchanging and eternal. Just as physical objects and phenomena evoke mathematical relationships in our mind, things with inherent beauty—a sight, a sound, or a fragrance—evoke an emotional perception of beauty in our soul, and the more intense that evocation, the more beautiful something is perceived to be.

Plato's theories of Form and Beauty have greatly influenced western culture throughout the ages, and, in particular, in the arts and theories of art. Some art scholars have seized upon Plato's ideas and theorized that true art somehow captures the ideal Form of Beauty inherent in real-life subjects. Others have linked beauty with spirituality and proposed a theory of *art by divine inspiration*, which associates Platonic beauty with truth or divine perfection. Such theorists postulate that beautiful works of art have a depth and subtlety of expression that enables them to transmit the artists' experience of the sublime or the divine. Michelangelo's sculptures and paintings have been cited as good examples of art by divine inspiration, and anyone who has visited the Sistine Chapel in the Vatican would likely be inclined to concur. Michelangelo himself has been quoted as saying, "The true work of art is but a shadow of the divine perfection."

Even before the time of Plato, ancient Egyptian and Greek artists strove to create paintings and sculptures that captured the inherent beauty in the human form. To this end, they developed *canons of proportionality*, which were sets of precise dimensions of underlying ideal archetypes to be used in the depiction of the human body. Early Egyptian papyrus scrolls and hieroglyphic carvings depict Egyptian gods, pharaohs, and common labourers all carefully drawn or crafted in accordance with the Egyptian canons of proportionality. And late classical Greek sculptures present gods and heroes as having bodies of ideal Form, created in perfect proportions and filled with a calm repose as if they inhabited a perfect and changeless divine world.

About 120 BCE, the Greek sculptor Alexandros of Antioch carved the famous statue *Venus de Milo*, considered to be amongst the most beautiful sculptures of the female body ever created. The Venus de Milo statue now resides in the Louvre museum in Paris and has enthralled admirers from all over the world since it was unearthed in 1820. It is believed to represent Aphrodite, the Greek goddess of love and beauty, and it certainly seems to capture Plato's Form of Beauty as manifest in the female form. In a way, the statue represents the culmination of humankind's attempts to capture the beauty in the female form that began with the Venus of Berekhat Ram, carved more than 250 thousand years ago.

The *Venus de Milo*.

During the first century BCE, the Roman architect and civil engineer Marcus Vitruvius Pollio wrote extensively about the perfect proportions of the human body and how they could be extrapolated to create pleasing architectural designs. Vitruvius's canons of proportionality were employed throughout the Roman Empire in the construction of buildings and in the sculptures that adorned them.

In 1490, Leonardo da Vinci embraced and improved upon the ideas of Marcus Vitruvius Pollio and produced his famous drawing *Vitruvian Man*. The drawing illustrates how the beauty inherent in the ideal human form can be projected to produce the most pleasing proportions for a cathedral built in the shape of a crucifix, the shape traditionally adopted for cathedrals throughout Europe during the Medieval and Renaissance periods.

Over the ages, most gifted sculptors and painters had canons of proportionality that they used in the design and crafting of their work. Scattered throughout the city of Rome, for example, Gianlorenzo Bernini's sculptures that adorn the Villa Borghese, the Sant'Angelo Bridge, and Saint Peter's Square seem to capture the very essence of the ideal human form and are truly remarkable in their beauty and grace.

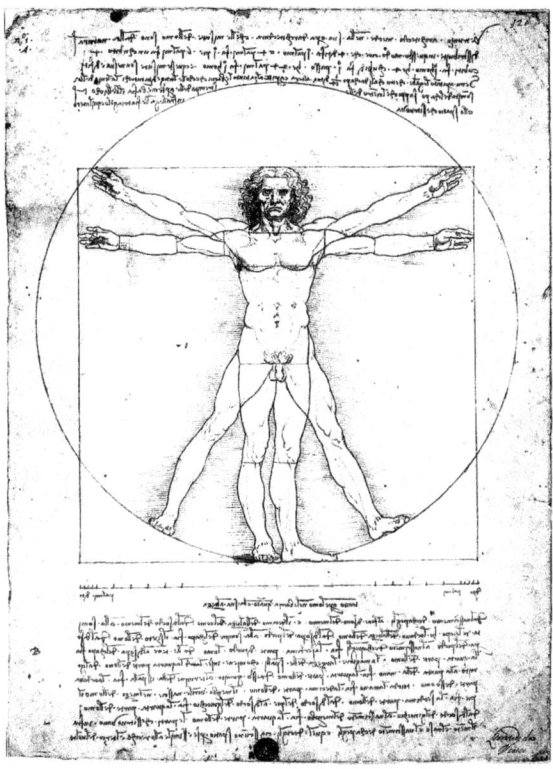

Leonardo de Vinci's *Vitruvian Man*.

Plato's eternal Form of Beauty, as well as his musings on the fleeting nature of physical beauty, has also influenced many poets and writers. Plato himself was something of a poet in his youth and understood the power of poetry to inspire and motivate people to act, but as he matured, philosophy came to appeal to him more as a means to discover truth and beauty. Somewhat ironically, in his later years, he became disdainful of poets and poetry, claiming them to be amoral with the inherent capacity to mislead. In *The Republic,* Plato supports poets and poetry only so far as they inspire others to perform good deeds and live a moral and ethical life. Nevertheless, Plato's musings on beauty as an ideal Form inspired several poets during the European Romantic period, including John Keats and William Wordsworth.

In the early part of the nineteenth century, the English poet John Keats was inspired by the art and literature of classical Greece and in particular by Plato's musings on beauty and art. In spite of Plato's somewhat negative musings about poets, Keats chose to write about the beauty to be found in ancient Greek sculptures and paintings and how their inherent beauty was timeless, providing enjoyment for all throughout the ages.

In his famous poem *Ode on a Grecian Urn*, Keats appears to embrace Plato's concept of beauty as an underlying perfect Form when he equates the inherent beauty of the urn with truth. And while we humans waste away into old age and lose the beauty of youth, the beauty represented by the urn endures, to be enjoyed by humankind for all eternity.

Keats summarizes these sentiments at the end of his poem:

> *"When old age shall this generation waste,*
> *Thou shalt remain, in midst of other woe*
> *Than ours, a friend to man, to whom thou say'st,*
> *'Beauty is truth, truth beauty'—that is all*
> *Ye know on earth, and all ye need to know."*

And along similar lines, the English poet William Wordsworth laments over the loss of the beauty of youth in his famous poem, *Ode on Intimations of Immortality from Recollections of Early Childhood,* with the inspiring words:

> *"What though the radiance which was once so bright*
> *Be now for ever taken from my sight*
> *Though nothing can bring back the hour*
> *Of splendour in the grass, of glory in the flower*
> *We will grieve not, rather find*
> *Strength in what remains behind"*

These words and phrases, like the writings of so many great poets and playwrights, seem to resonate with our emotions and evoke the perception of Plato's Form of Beauty in our soul.

In the world of sound, music represents conceptualization in its purest form. Of all the arts, music is unique in its ability to stir our emotions, sooth the troubled soul, awaken our humanity, give us hope in times of despair and courage when faced with adversity. Even the simplest of melodies seems to stir our minds and bodies in such a way that we often can't resist the urge to tap our feet or dance and sway to the beat of the music, sometimes even without thinking.

Music resonates with our soul and strengthens our spirituality, often playing a central role in worship. The great cathedrals of Europe have large, powerful pipe organs ornately integrated into their interiors, and many of the great classical composers of history wrote organ music to be played in the religious ceremonies conducted there. Ludwig van Beethoven, one of the most admired composers of such music, is purported to have once said, "Don't only practice your art, but force your way into its secrets, for it and knowledge can raise men to the Divine." No one can deny that the sound of organ music resonating through the nave of a beautiful cathedral is a truly inspiring experience, lifting one's soul out of the stresses and trials of everyday life and enabling one to rejoice in the presence of the creator, regardless of what one perceives that creator to be.

Great music has a timeless appeal. Some music has such inherent beauty that it can carry our emotions to extremes of sorrow, nostalgia, joy, and ecstasy, and no matter how many times we hear some musical pieces, our souls continue to be transported to these exalted realms of emotional intensity.

The beauty in music is felt and enjoyed by everyone, regardless of their spoken language or cultural background. There is a primeval appeal about music, and scientists speculate that the production and enjoyment of music was one of humankind's earliest forms of cultural expression, likely arising in concert with the first developments of language, if not even before that.

And so, it seems that while mathematics is the universal language of the mind, music is the universal language of the soul.

Our Astonishing Ingenuity

> *"The intuitive mind is a sacred gift and the rational mind is a faithful servant. We have created a society that honours the servant and has forgotten the gift."*
>
> ...Albert Einstein

Human ingenuity, that extraordinary capacity for reasoning and creativity that each of us possesses, is a hallmark of our species and the principal factor accounting for our astounding evolutionary success. It is our ingenuity that enables us to acquire scientific knowledge, fabricate technological marvels, create beautiful works of art, and formulate our spiritual beliefs. How our species came to acquire these capabilities has to be one of the greatest mysteries of evolution.

The first indications of human ingenuity emerged about two million years ago when *Homo habilis* began to craft Oldowan butchering tools to help satisfy their developing taste for meat and marrow. Ever since that first spark of ingenuity appeared, our human ancestors have continued to produce an ever-increasing variety of innovative devices and solutions to facilitate their daily routine, improve their survivability, and enhance their quality of life. And this proliferation of products of our ingenuity has continued unabated into modern times.

Today we are surrounded by a staggering array of technological tools and devices that have added enjoyment and comfort in all aspects of our lives. Such innovations have revolutionized our cities and homes, our transportation and communication systems, and our cultures.

Our appetite for the faster and better appears to be insatiable, and the rate at which we continue to innovate is still increasing exponentially. Just when we think we have created the ultimate machine or device that couldn't possibly be improved upon, a new one is invented that is even better and faster. No one can predict the things that future generations will create, but if there is one thing of which we can be certain, it is that there will be technological marvels with capabilities that we can scarcely even imagine today.

What is it about our minds that makes us such prolific innovators?

Human creativity stems from a combination of mental abilities that starts with conceptualization and includes reasoning, intuition, and imagination. When such capabilities are combined with our insatiable curiosity and a compulsion to continuously improve upon the circumstances in which we find ourselves, the result is that every individual in our society is constantly innovating to some degree. Humans thrive on change and are constantly looking for ways to make tasks easier or to do them better, to create or acquire new things. It is this driving propensity for the different and better that continuously improves our quality of life, and since the dawn of civilization, it has fuelled our economies as well.

Intuition and imagination are the key elements underlying this incredible creativity.

Intuition is that somewhat mysterious ability that humans have to recognize relationships in factual information, to connect the dots as it were, and to form models of the behaviour of natural phenomena. Our intuition enables us to form ideas and identify relationships amongst concepts that, in turn, enable us to interpret and understand the world around us. Intuition is that process in the human mind whereby ideas spring forth, problems are solved, and decisions are made.

Typically, when attempting to solve a problem, we think about the facts and information available in our memory, and apply logic to find a common thread or element in the information that will reveal a solution. It is our intuition that accounts for that sudden spark of realization that comes when we finally find that solution.

Intuition plays the same role in all of our thinking, from mundane activities—such as identifying a word in a crossword puzzle or remembering where we misplaced a personal item—to the sublime—composing a melodic theme for a new concerto or devising a new theory like quantum mechanics to account for the behaviour of subatomic particles. All humans possess intuition to a varying degree, and it is often described as the sixth sense, after the five faculties of sight, hearing, touch, smell, and taste.

Closely aligned with intuition is the uniquely human capacity for imagination, that remarkable ability to conceive and visualize things that don't exist. Intuition and imagination are not the same, although they tend to be inextricably linked. Those with a strong intuition typically have a fertile imagination as well.

While intuition tells us what is or must be true, our imagination tells us what might or could be true. Unlike our intuition, our imagination is not bound by reality and can operate with equal ease in a fictional, imaginary world or a tangible one. It is our imagination and intuition, working in concert, that enable us to write science fiction stories, develop new scientific theorems, and invent new machines and devices.

Intuition is the key to acquiring knowledge, whereas imagination is the key to creativity. Our exceptional creativity and innovation are what fuel our cultural development. They are the main reason why *Homo sapiens sapiens* is the only species to have ever experienced a cultural explosion such as ours.

When an intuitive idea occurs, it typically comes in a *flash*, sometimes without conscious proof or supporting evidence. But we are instantly certain of its truth, and even though we can't explain why we know it is true, we *just know* it is. This *truth conviction* is critical in fields like scientific research because it can sometimes take years for a scientist to prove a theory mathematically after having an initial breakthrough idea that may have occurred to them in a few seconds. Only a strong conviction that an intuitive idea must be true could provide the motivation to spend years trying to prove it.

The American inventor and businessman, Thomas Edison, has been credited with the invention of the phonograph, the motion picture camera, and the electric light bulb, plus many other ground-breaking technological devices, and he is considered by many to be America's greatest inventor. When asked to explain the source of his genius, Edison remarked, "Genius is 1 percent inspiration and 99 percent perspiration." The inspiration was that intuitive flash that gave him an idea for a new invention, and the perspiration was all that hard work required to translate it into a useful device.

Every human has intuitive capabilities to a greater or lesser degree, and most of us can recount experiences when our intuition has helped us solve a problem, develop a course of action, or increase our understanding of something. Some people feel uncomfortable acting on intuitive ideas, while others rely on their intuition as an important element in the conduct of their daily lives. Successful people in virtually all walks of life, including music, the arts, business, politics, science, and engineering, rely heavily on their intuition and imagination.

Our intuition does not always come through for us, at least not immediately. Sometimes, we can think about a problem for a long time and fail to come up with a solution. Our intuition appears to fail us and eventually we give up and move on to think about something else.

But it seems that our intuition doesn't stop there. Somehow, our mind keeps working the problem in the background, as it were, and suddenly a solution pops into our mind out of the blue, even when we are thinking about something else. Some people who rely on creativity to make a living report that if they think about a problem and fail to solve it before retiring for the night, the solution sometimes springs to mind while they sleep, either waking them up during the night or being present in their mind when they awaken the next day. As a result of such phenomena, we speak of a subconscious mind as having the ability to ponder problems independently of our conscious mind, continuing to work on a problem even when our conscious mind has moved on to something else or shut down completely during sleep.

Intuition and instinct are sometimes held to be the same thing, particularly when it comes to guiding behaviour or taking certain action in the face of complex situations. But they are, in fact, very different. Whereas instinctive behaviour involves no direct conscious thought, intuition is very much an activity of the conscious mind, a conscious experience whereby we suddenly know that something is true or that we need to act in a certain way. We often can't explain where the idea came from, but it does occur in our conscious mind and requires thought to translate the idea into action. When we act or behave *instinctively*, there is no initiating conscious thought involved. We just react without thinking, often suddenly or impulsively before we've even had time to think about it.

It must be admitted, however, that it can sometimes be difficult to tell whether we are acting instinctively or intuitively. Occasionally, when we feel strongly inclined to proceed with a certain course of action, it can be our intuition that is guiding us even though we are unaware of the reasons for feeling so compelled, or it may be an instinctive urge that comes from a genetic predisposition that was encoded in our genes many millions of years ago by the evolutionary process.

Ingenuity and Spirituality

"We are not human beings having a spiritual experience. We are spiritual beings having a human experience."

...Pierre Teilhard de Chardin

Humans are deeply curious beings with an insatiable thirst for knowledge and understanding about everything around us. It seems we all possess an inherent need to learn everything there is to know about the natural world in which we live, the universe in which our Planet Earth is situated, and the nature and meaning of life itself. Knowledge about why natural events and phenomena behave as they do provides us with an explanation for them, and to have an explanation for something is to understand it. Knowledge and understanding are deeply satisfying, providing us with comfort and security, somehow making it easier, both intellectually and emotionally, to accept and endure hardship or adversity when it comes. We are equally disturbed, sometimes even frightened, when something happens for which an explanation eludes us.

This intense desire for knowledge applies to the scientific and to the spiritual. Since time immemorial, humankind has hungered for answers to the enduring questions of antiquity: Where do we come from? What is the meaning of life? Is there such a thing as a soul or god?

Early *Homo sapiens sapiens* would have had the same inherent thirst for knowledge as we possess today. Even in the very earliest human societies, scientific understanding and spiritual belief would have developed in concert, each influencing the other to form an integrated world view, just as they do today.

Material or scientific knowledge and spiritual belief are the foundations of human culture. Scientific knowledge enhances our survival both as individuals and as a species, while our spirituality enables us to enjoy our life experience. Together, they harmonize to enrich our lives and allow us to appreciate life as the truly wonderful gift that it is, enabling and motivating us to live full, enjoyable, and meaningful lives.

Our cultures are strongly influenced by our spirituality. Artistic expression often reflects religious belief as any visitor to the Louvre museum in Paris or the Vatican in Rome can testify. In Western societies today, our customs, laws, and social organization are largely based on Judeo-Christian values. Our *work week* is based on the creation myth of the Book of Genesis, including keeping Sunday as a day of rest. In Christian societies, the two major holidays, Christmas and Easter, celebrate the birth and resurrection of Jesus Christ, and the very term holiday is an abbreviation of holy days. Western societies' laws and ethics are largely based on the Ten Commandments, which were passed from God to Moses as described in the Book of Exodus in both the Torah and the Old Testament of the Christian Bible, and on the teachings of Christ found in the New Testament of the Christian Bible.

Humans apply logical reasoning, intuition, and imagination in the advancement of both our scientific knowledge and our spiritual beliefs.

We acquire knowledge through a process of observation and reflection. When we observe the world around us through our five senses, our brains extract information from the sensory data and store it in our memory. We think about the information received and seek to identify underlying truths and relationships within it. Then our intuition springs into action, formulating concepts, ideas, and theories to capture those truths or account for those relationships. Those concepts and theories constitute knowledge, enabling us to explain and predict the behaviour of natural phenomena.

Most of the knowledge humans acquire is about the material world, and coupled with our creative imagination, such material or scientific knowledge provides the foundation for scientific research, technology development, and the fabrication of utensils, tools, weapons, and devices. Such inventions facilitate the performance of our daily routine, relieving us from onerous or tedious tasks and freeing up more time to spend on those things that we enjoy most. Scientific knowledge also affords us an understanding of the workings of the universe and provides us with insight into how the natural world functions, including how living organisms grow, function, and interrelate. Such knowledge, in turn, provides us with the satisfaction that comes with understanding and enables us to prolong our lives through the prevention and curing of disease.

Material or scientific is not the only type of knowledge our intuition provides, however. Our intuition also has the capacity to acquire spiritual knowledge.

When we observe nature, we often see things such as the symbiotic relationship between the bee and cherry blossom, which reminds us of the wonderful web of interdependence that exists amongst all living things. Such observations, when combined with our scientific knowledge of the underlying principles that govern the functioning of the cosmos, inspire us to appreciate the infinite wisdom underlying the behaviour of the natural world and to ponder the source of that wisdom. On such occasions, our intuition tells us that there must be a supreme consciousness, which is the source of the wisdom inherent in the cosmos.

When we witness the birth of a child or the death of a loved one, we reflect on the nature of life's vitality and wonder about its source. Our intuition tells us that there must be a universal source of the vitality of life that animates an organism at the time of conception or birth and that leaves the organism at the time of death.

The process whereby scientific or material knowledge is acquired has been formalized into what is known as the *scientific method*, a rigorous procedure whereby a theory is postulated and then experimentation is performed to either confirm or refute the hypothesis. If the theory withstands the scrutiny of experimental confirmation by several independent scientific investigators, it becomes a scientific law and is accepted as scientific knowledge or fact.

We apply a similar process in the acquisition of spiritual knowledge. When we have a strong emotional or spiritual reaction to a particular life experience or entertain intuitive thoughts and ideas about the spiritual aspects of existence, we find that others have the same or similar spiritual ideas or theories in response to the same experience. Such affirmation constitutes the equivalent of the scientific method applied to the acquisition of spiritual knowledge—in effect, a *spiritual method*. In reality, there is little difference between the two processes or their validity. The spiritual method applies the same methodology to verify our spiritual theories and ideas as the scientific method applies to our theories and ideas about materiality. Both rely on the same intuition to formulate those initial concepts, and both rely on confirmation and acceptance by our peers before being accepted as fact or belief.

Our intuition simply cannot accept the possibility that the cosmos is a random, meaningless occurrence or that the vitality that is life simply appears from nowhere at conception and evaporates into nothingness at death. Neither the cosmos nor life can be just a capricious *something* that suddenly appears out of *nothing*, as it were. Such a notion not only runs counter to our intuition, it seems illogical.

For most of us, our intuition strongly supports the notion that the beauty of the natural world and the wisdom inherent in the laws that underlie the workings of the cosmos, imply the existence of a supreme consciousness, the creator of the universe. And so too, life's vitality can only be the manifestation of that supreme consciousness within all living things. Such notions constitute our spiritual beliefs or perhaps, more appropriately, our spiritual knowledge. Such beliefs or knowledge are central to all of the world's great religions and are held by the vast majority of people living throughout the world today.

Most of us harbour a deep conviction that the universe was created for a reason and that our lives have meaning and purpose. We intuitively feel that Planet Earth, with its incredible diversity of flora, fauna, and natural wonders, is simply too awesome and that life is simply too wonderful for it all to be the result of a series of meaningless, random events.

But since the beginning of the scientific revolution, humankind has come to place much more credence on scientific knowledge than spiritual, dismissing the latter as fantasy and referring to it as unfounded superstition. Today, many of us trust our intuition to acquire scientific knowledge implicitly, but we are reluctant to trust that same intuition when it comes to acquiring spiritual knowledge.

As scientific discoveries have deepened our understanding of the laws and principles that govern the behaviour of the universe, some of us are inclined to interpret such breakthrough discoveries as proof that the universe operates in a perfectly explainable manner, requiring no further rationale or explanation as to its meaning or origins. Such sentiments are known as *naturalism*, the philosophical belief that everything arises from natural properties and causes, and that supernatural or spiritual explanations are both unnecessary and unfounded.

Naturalism has its roots in the *Theory of Scientific Determinism*, which dates back to 1687 when Isaac Newton published his monumental work, *Philosophiae Naturalis Principia Mathematica*, where he laid out the mathematical principles of time, force, and motion that remain the foundation of the physical sciences and engineering today. Newton's laws of motion and universal gravitation were such a giant leap forward in the understanding of the mechanistic behaviour of the natural world that a certain hubris beset the scientific and philosophical communities, which led to the belief that the entire universe was unfolding in accordance with mechanical principles that could be understood and that would allow us to one day predict the future characteristics and behaviour of every particle of matter in the universe.

Scientific determinism held that everything observed in the universe had a physical explanation, that nothing was left to chance. The future was deterministic or predictable by scientific laws, some of which were yet to be discovered but inevitably would be. By implication, free will was an illusion, belief in the existence of a soul or any other non-material entity was baseless, and the future was predetermined. Individuals were not responsible for their behaviour since they were powerless to change it. God, if He existed at all, was relegated to the role of initiator of the universe and, after that, became only an observer of how it unfolded. Scientific determinism reinforced a monistic view of consciousness as having only a physical manifestation that would one day be discovered and be found to be explained by scientific laws and principles.

Fortunately for the world, scientific determinism largely remained academic conjecture, and few people actually subscribed to the theory, at least to the degree where it affected their behaviour. The theory was also relatively short-lived because around the beginning of the twentieth century, scientists began to unravel the secrets of the atom. When they were finally able to observe the behaviour of subatomic particles, it was discovered that not only did they not behave in accordance with Newton's laws of motion, their behaviour was rather unpredictable and could only be characterized statistically. Given that the basic building blocks of all matter had an element of randomness to their behaviour, by implication, so too did the macro elements of the universe they composed, and the future course of natural phenomena could not be known precisely after all. Free will had been reinstated; our future was not entirely predictable, and the existence of a soul and God were once again possibilities.

For naturalists, scientific knowledge seems to refute the notion of a supreme being or God, but for others, scientific understanding only adds to the grandeur and mystery of the universe in which we live. Many well-known scientists, including Albert Einstein, Stephen Hawking, and Nikola Tesla, have admitted that the laws governing the behaviour of the universe reflect such a profound wisdom that they simply could not accept the notion that it is the result of a random, meaningless event. Even Isaac Newton, the very man whose monumental work led others to postulate the theory of scientific determinism, once confessed, "This most beautiful system of the sun, planets, and comets, could only proceed from the counsel and dominion of an intelligent and powerful Being."

Ingenuity and Survival

"As a species, we've somehow survived large and small ice ages, genetic bottlenecks, plagues, world wars and all manner of natural disasters, but I sometimes wonder if we'll survive our own ingenuity."

...Diane Ackerman

Until modern times, human ingenuity has operated with impunity, consuming the earth's resources and modifying the earth's environments at an ever-increasing rate, creating ever more numerous and diverse technological devices. At the same time, the numbers of *Homo sapiens sapiens* on earth continue to grow exponentially as we apply our ingenuity to the extension of our lifespan by addressing the traditional limiters of starvation, predation, injury, and disease.

But while our technological and cultural achievements do much to enrich our lives, it remains uncertain, from an evolutionary perspective, whether they further enhance our ability to survive as a species.

The invention and production of ever more ingenious machines and devices has become an all-consuming passion in modern times, and their consumption by individuals, or consumers, has become the mainstay of our economic prosperity. The accumulation of wealth and material things has now become the defining goal of most human societies, and we have come to measure our individual success by the possessions we accumulate. And we measure a nation's success by the amount and value of the goods it produces and the services it provides, known as its Gross Domestic Product or GDP.

As a result, humans have become the scourge of the planet. We are exhausting the earth's capacity to provide us with the resources we need, and we are wreaking irreparable damage on the earth's environments as we continue our relentless drive to increase the GDP of our societies and greater prosperity for ourselves as individuals.

Humankind is now faced with the prospect that our remarkable powers of creativity and innovation may simply be too powerful for the earth to sustain. Today, the pace of our innovation continues to accelerate, seemingly without limit, and we have recently reached the point where the by-products of our economic development are contaminating the earth with toxic chemicals and discarded waste at such a rate that many of earth's ecosystems and biomes have been damaged beyond recovery. Most significant in this regard is the enormous increase in the production of greenhouse gases, which are causing the earth's atmosphere to warm at an accelerating rate, giving rise to massive climate and ecological changes throughout the planet.

We must never forget, however, that nature is master of her own ship and owes *Homo sapiens sapiens* no favours. If we don't soon temper our lust for economic growth and apply our ingenuity towards working with nature to establish a sustainable balance between our prosperity and the natural environmental restorative processes, nature herself will surely intervene, and her solutions will not be kind to *Homo sapiens sapiens*.

As the eminent astronomer and cosmologist Carl Sagan once commented, "Extinction is the rule. Survival is the exception."

Left untempered, Modern Human behaviour could very well lead to the drastic depopulation, if not outright extinction, of our species, along with most other plant and animal species on the earth. If we let this happen, rather than enhancing the survival of our species, our "advanced" cognition would be responsible for yet another mass extinction, and the remarkable mind of *Homo sapiens sapiens* would prove to have been the most spectacular evolutionary failure the earth has ever witnessed.

CHAPTER THIRTEEN
Consciousness, Vitality, and the Soul

"Looking for consciousness in the brain is like looking inside a radio for the announcer."

...Nassim Haramein

OF ALL THE amazing capabilities of the human mind, consciousness itself is arguably the most fascinating, yet in spite of the vast amount of research that has gone into understanding its physical basis, the source and nature of consciousness remains a mystery. Consciousness appears to be simply a property of being alive, a state of awareness that gives us the ability to have thoughts, form perceptions, and experience emotions. It is something that all living entities appear to possess to a greater or lesser degree, and it is the mechanism by which living organisms experience life.

One of the fundamental elements of consciousness is self-awareness, which appears to be demonstrated by even the lowliest of life forms. Even single-celled organisms can sense their surroundings and determine if objects close-by provide food or pose a threat, reacting accordingly. While such behaviour is purely instinctive, it can be argued that it demonstrates at least some level of an awareness of self as distinct from something separate from the self, which represents a rudimentary form of consciousness as well.

Consciousness seems inextricably entwined with the vitality of life. All living organisms possess a certain *vitality*, an animating energy, or *life force*, that imbues an organism with the compulsion to survive, flourish, and embrace life.

To a biologist, an organism is defined to be alive, or *animate*, if it is self-contained, physically distinct from its surrounding environment, responsive to external stimuli, and capable of the metabolic processes necessary for self-sustainability and reproduction. Hence, life or vitality is defined as the ability of an organism to exhibit these capabilities—an ability that clusters of organic molecules somehow spontaneously acquire after they have formed into a physical configuration that is capable of sustaining and reproducing itself. Like consciousness, the vitality of life can only be defined as an *ability* as it has no physical attributes in itself, and its source, where it comes from at birth and where it goes at death, is a mystery.

However, an organism having the ability to do something does not mean that it will, in fact, do it. The vitality of life also instills an organism with a propensity to be proactive, to actually do the things it is capable of doing to ensure its survival, growth, and reproduction. Vitality provides that mysterious inner impulse that motivates us to keep living in the face of the most severe hardships and to fiercely resist death when faced with the most daunting threats to our lives. This compulsion is known to an evolutionary biologist as the *survival instinct*, and from an evolutionary perspective, it is what compels all living organisms, whether bacteria, plant, or animal, to relentlessly act in a manner that ensures their survival and maximizes their chances to reproduce.

The compulsion to live and reproduce is particularly strong in more complex life forms. All animals use every means they have to feed themselves, to prolong their lives as long as possible, and to reproduce, but the more complex life forms will go to extreme behaviours and endure unimaginable adversity to find sustenance and avoid death. Whether they are searching for food, fighting for their lives, or fleeing from mortal danger, all animals will push themselves beyond what their bodies can sustain—ultimately, to the point at which their body collapses, and they succumb to exhaustion.

As unlikely as it seems, even single-celled, microscopic life forms possess some rudimentary life force that drives them to avoid threats to their existence, to sustain themselves, and to reproduce. Most bacteria are capable of moving purposefully, gathering in regions that are hot or cold, that are light or dark, or that contain nutrients. They can sense their environment, seek safer and more hospitable surroundings, and hunt for food. The physical means to perform these tasks is evident in their cellular structure, and their behaviour is entirely instinctive, but where this *drive to survive* comes from, is unknown, at least to science.

The existence of a life force, that intense compulsion to survive and reproduce, fuels the evolutionary process and underlies the mechanism of survival of the fittest. The compulsion to live, flourish, and reproduce is what motivates an organism to compete with other organisms for survival, using every advantage that evolution has afforded them. This is true for all life forms, whether an amoeba, a rhododendron, or a blue whale.

But vitality goes beyond providing the compulsion for individual organisms to survive and reproduce. Life, it seems, has a compulsion to simply exist in as many physical, organic forms as possible and in any and every environment to which it can possibly adapt. Within a very short time of life's first appearance, the earth's oceans were teeming with microscopic life, numbering in the millions of species and trillions of individuals. During the Cambrian Explosion, countless numbers of new species were suddenly swimming throughout earth's oceans, and when life finally gained a foothold on land, another explosion of life forms took place. Within a short evolutionary time frame, millions of species of plants, animals, fungi, and bacteria populated every square millimetre of the land, adapting to virtually every environment that Earth had to offer. Today, the number of individual life forms occupying the earth's air, land, and seas is so large that it is impossible to enumerate or even contemplate. A single spoonful of organic garden soil may contain as many as ten thousand *species* of bacteria and a billion individual life forms.

The mechanism by which complex organic molecules transition to living cells is still unknown to science. Although scientists have reproduced many of the intermediate steps in the formation of complex organic molecules, they have not yet succeeded in creating living cells in the laboratory. However, life somehow arises spontaneously in complex organic molecular structures whenever and wherever the conditions are suitable and sustainable. Inevitably, therefore, scientists will one day succeed at replicating those conditions, and life will spontaneously appear in synthesized, complex, organic compounds. But this does not mean that science will have created life. It *will* mean that science has succeeded in creating a complex organic molecular structure that is capable of hosting and sustaining life. The true nature of life will remain a mystery, however, even after scientists have succeeded in bringing it into existence in a Petri dish, which, after all, is just another environment in which the vitality of life can achieve a physical existence, which it seems, it is compelled to do.

Consciousness and vitality have long been associated with an inner spirit, or *soul*, that most ancient and modern cultures deem to reside in all living things. Each of the three entities: consciousness, vitality, and the soul, defy human understanding in any physical sense, and yet their connectedness is widely acknowledged. And while consciousness is often associated with the brain, and vitality with the heart or body, consciousness, vitality, and the soul are, in most contexts, inseparable and indistinguishable.

The Birth of Theism

"God reveals himself in the orderly harmony of what exists."

...Albert Einstein

Prior to the advent of agriculture, twelve to fourteen thousand years ago, human societies were organized into small hunter-gatherer bands, typically limited by available resources to less than one hundred individuals. If band size grew too large for the available resources to support, they would divide into separate bands that relocated some distance apart.

Virtually all hunter-gatherer societies practiced some form of animism, a belief that a supernatural power, or *Great Spirit*, organizes and animates the material universe and that this great spirit is manifest in all living entities as an inner spirit. In many animistic cultures, geographical features such as mountains, rivers and lakes, and even the elements—the wind, the clouds, and thunderstorms—were also deemed to possess a distinct spiritual essence.

With the advent of agriculture in places like Mesopotamia, China, and India, our nomadic ancestors began to adopt a more sedentary lifestyle, as they could now grow enough food to support themselves in one area. Before long, settlements arose, and people no longer wandered very far as they needed to stay close to home to tend their crops and care for livestock. As their agricultural technology improved, they could support more people, and group size increased. Bands grew into tribes, and tribes grew into ever-larger communities. Encampments became villages, then towns, and ultimately cities. Societies grew from a few dozen to several hundred people, then thousands. By about 3000 BCE, populations of cities throughout Mesopotamia and the Fertile Crescent were numbering in the tens of thousands.

As community size grew, the need for more social structure developed. People needed to be assigned to specific tasks related to growing crops and tending animals. The agriculture-based communities were most frequently located in river valleys where the climate was favourable and there was an abundant source of water for the crops and livestock. To effectively harness the water, irrigation and terracing became necessary, and the work of constructing such things required a work force that had to be organized and managed. Political control over large areas where such infrastructure was being built became necessary. These improvements in agriculture led to a surplus of food and other resources, which meant that wealth was accumulated, typically by those in control. Before long, communities became subdivided into socio-economic classes.

A hierarchy of authority, based upon class, inevitably evolved as the need for rulers, rule enforcement, and compliant workers emerged. New rules or laws had to be established to provide the framework for a new code of behaviour that would preserve order and harmony across the developing socio-economic class structure.

Around this time, the earliest forms of polytheistic religion appeared as some of the spirits of animistic beliefs were elevated to the status of deities. These gods were capricious, and many were distinctly ungodlike, with much the same character attributes and flaws as humans. They were deemed to be powerful and judgmental however, demanding that humans pay homage and tribute to them and behave in certain ways that would appease them. The moral code of behaviour that the socio-economic structure required was reinforced through deities' threats of punishment in the form of crop failure or natural disasters such as flood, storms, fire, and disease.

Societies in Mesopotamia, Greece, Rome, and Egypt developed formal pantheons of such gods, and a new social class of priests and priestesses evolved to act as intermediators between the gods and the people. Only the priests and priestesses could communicate with the gods to learn what was required to appease them, and these demands were passed on to the rulers and the people at large. Understandably, the priests and priestesses became extremely powerful, obtaining a status that often rivalled that of the rulers.

It was against this religious and political landscape that the seeds of modern, monotheistic religion were sown. Rejecting the notion that capricious gods were manipulating natural phenomena to torment humankind, some ancient Mesopotamian sages began observing the natural world for what it truly was, and they came to realize that the cosmos was operating in an orderly fashion and in accordance with certain fundamental principles. They learned to appreciate the unfathomable wisdom inherent in those principles and reasoned that such wisdom could only stem from the supreme consciousness, or mind, of an omniscient creator. They developed a reverence for this divine consciousness and declared it to be the one and only true God. Unlike the traditional gods, they believed the true God to be fair and just, and by about two thousand years BCE, a devoted group of people emerged, known as Israelites, that rejected the traditional pantheon of gods and began to worship the one true God.

The ancient Israelites believed they had a unique covenant with God, whereby God would guide and look after them provided they strived to conform to God's will and live a good and moral life. And as time passed and their numbers grew, they accumulated an oral tradition of myths and legends that recounted the history of their relationship with God in the context of that covenant. The history of the Jewish people, as they are now called, describes the trials and tribulations they endured as they strove to honour their covenant with God and follow a way of life embodied in the set of moral, religious, and civil obligations deemed to have been given them by God.

And it was thus that the world's first monotheistic religion, Judaism, was born.

Between 700 and 300 years BCE, scribes in Babylonia documented the Jewish tradition on clay tablets and scrolls of papyrus and parchment. These writings have survived to modern times and constitute the first five books of the Hebrew Bible —Genesis, Exodus, Leviticus, Numbers, and Deuteronomy, collectively known as the Torah.

The first book of the Torah, Genesis, opens with an account of the creation of the cosmos, which begins:

> *"1:1 In the beginning God created the heaven and the earth.*
>
> *1:2 And the earth was without form, and void; and darkness was upon the face of the deep. And the Spirit of God moved upon the face of the waters.*
>
> *1:3 And God said, Let there be light: and there was light.*
>
> *1:4 And God saw the light, that it was good: and God divided the light from the darkness.*
>
> *1:5 And God called the light Day, and the darkness he called Night. And the evening and the morning were the first day.*
>
> *1:6 And God said, Let there be a firmament in the midst of the waters, and let it divide the waters from the waters.*
>
> *1:7 And God made the firmament, and divided the waters which were under the firmament from the waters which were above the firmament: and it was so.*
>
> *1:8 And God called the firmament Heaven. And the evening and the morning were the second day.,*
>
> *1:9 And God said, Let the waters under the heaven be gathered together unto one place, and let the dry land appear: and it was so.*

> *1:10 And God called the dry land Earth; and the gathering together of the waters called he Seas: and God saw that it was good.*
>
> *1:11 And God said, Let the earth bring forth grass, the herb yielding seed, and the fruit tree yielding fruit after his kind, whose seed is in itself, upon the earth: and it was so.*
>
> *1:12 And the earth brought forth grass, and herb yielding seed after his kind, and the tree yielding fruit, whose seed was in itself, after his kind: and God saw that it was good.*
>
> *1:13 And the evening and the morning were the third day."*

The narrative then goes on to provide a step-by-step description of how, over the course of six days, God created the cosmos, including the earth and all life forms, culminating with the creation of humankind as described in Genesis 2:7, "*And the Lord God formed a man from the dust of the ground, and breathed into his nostrils the breath of life, and the man became a living soul.*" Hence, in Judaism, it was God that created the cosmos and imparted vitality, consciousness, and a soul to humankind.

The Genesis creation narrative was first written almost three thousand years ago, before the adoption of the scientific method as a disciplined approach to understanding the natural world. At the time, nothing was known about the age of the universe or of the natural processes upon which it operates. For example, the narrative says the cosmos was created in six days, which, in the light of today's scientific understanding, is certainly not the literal case. However, as our scientific knowledge of creation and the evolution of life has advanced, so too has our concept of historical time. Our cosmological theories now put the age of the universe at just under fourteen billion years and the age of the earth at four billion years, whereas in ancient Mesopotamia, the earth was believed to be about ten thousand years old. In fact, the numerical terms *million* and *billion* were only first coined in the fifteenth century CE.

Time frames of millions or billions of years are virtually impossible for the human mind to grasp, and hence, we tend to view the past in what mathematicians call a *logarithmic fashion*, where events about which we know very little and that occurred over large time frames in the distant past are compressed in our minds, while more recent historical events, about which we remember more detail, are less time-compressed, or even expanded. Recognizing that humankind's perception of time has evolved over time, some scholars have developed algorithms that translate biblical time frames into our modern perception of time. In their book, *The Biblical Clock,* authors Daniel Friedmann and Dania Sheldon have shown that when such techniques are applied to the Genesis narrative, the resulting time frames associated with the various creation events align quite well with current scientific theory. Considering this, and the state of humankind's scientific knowledge at the time it was written, the Genesis narrative provides an account of creation that is surprisingly consistent with modern scientific theory on the formation of the universe and the evolution of life on earth. The Genesis creation myth is indeed a testament to the wisdom and powerful intuition of the sages that conceived it.

For centuries, Judaism remained the sole monotheistic religion on earth. But the Hebrew Bible contains a reference to *the Messiah*, a saviour and liberator that will appear one day on earth to redeem the Jewish people. While today, the followers of Judaism still await the arrival of the Messiah, about two thousand years ago, a man named *Jesus of Nazareth* appeared in northern Israel that a few of their members at the time believed to be that Messiah. Jesus preached a new gospel and offered a new covenant whereby if an individual accepted his teachings, they were assured of salvation of their individual soul and would be granted an afterlife with God in heaven. The devoted disciples of Jesus came to believe him to be the son of God incarnate and to refer to him as Jesus Christ. The devoted followers of Christ and the covenant he taught, called their faith, *Christianity.*

Amongst the followers of Christ were a particularly devoted few known as the Twelve Apostles, who became his closest disciples during his life and ministry, and the primary teachers of his message after his death. Some of the apostles, along with other devoted disciples, wrote down the story of Jesus' life and teachings, and these now constitute the New Testament of the Christian Bible.

At the age of about thirty-three, Christ was crucified by Pontius Pilate, the Roman governor of the province of Judea, and the fledgling Christian community was persecuted by the Romans for over two centuries following his death. But still the community survived and continued to amass ever larger numbers of followers. Today, Judaism and Christianity remain separate faiths, and while Judaism is practiced by only a fraction of one percent of the world's population, Christianity, with almost thirty percent of the world's population as confessed adherents, is the dominant religion of the modern world.

The Concept of Universal Consciousness

"Seeing that the universe gives birth to beings that are animate and wise, should it not be considered animate and wise itself?"

...Zeno of Citium

At the time that the Mesopotamians were pursuing mysticism as a means to explain the mysteries of creation, sages in nearby Greece were adopting a different approach. By about 500 BCE, a few Greek philosophers had adopted a rigorous process of observation, experimentation, and inductive reasoning to understand the natural processes that were governing the operation of the cosmos. These *natural philosophers* as they are called, developed the world's first mathematical theorems and began to use them to describe the behaviour of material objects, producing the first scientific principles. Many consider this period to mark the beginning of the natural sciences and the origin of the scientific method, as some of the first scientific and mathematical theorems were conceived during this time. The mathematical and scientific principles established by Pythagoras, Euclid, Thales, and Archimedes, to name only a few, are still taught in secondary schools and universities today.

One of the earliest of these natural philosophers was Anaxagoras of Clazomenae. Anaxagoras was born around 500 BCE in Ionia, on the west coast of present-day Turkey. As a young man, he moved to Athens, Greece, in order to further pursue his interests in science and philosophy. His ideas influenced several of his contemporaries, including the statesman and politician Pericles and the playwrights Euripides and Aristophanes. A century or so later, his theories concerning creation and the composition of the universe influenced many of the great natural philosophers of classical Greece—Socrates, Plato, Aristotle, and Zeno.

Anaxagoras was a keen observer of the natural world and produced a number of novel scientific explanations of natural phenomena. He is credited with being the first to offer a correct explanation of eclipses, the interposition of another body between the earth and the sun or moon. He declared the sun to be a fiery mass of red-hot metal, larger than the Peloponnese; the moon to be composed of material similar to the earth; and the stars to be fiery stones. Considering that no body of scientific knowledge existed for Anaxagoras to draw upon, the insight demonstrated in his theories is remarkable as many of the fundamental concepts he postulated bear a strong resemblance to modern scientific theory and knowledge.

Anaxagoras is best known for his theories on the formation of the universe and the nature of matter. Like the Mesopotamian mystics, he observed that the universe was unfolding in accordance with a set of principles or laws, and he wondered about the source of those laws, eventually coming to the conclusion that they had to reside in some sort of universal mind or consciousness. Anaxagoras called this universal consciousness the *nous*, the Greek word for reason or mind. He reasoned that it was the nous that was responsible for the creation of the universe and the source of the laws governing its ongoing behaviour.

Anaxagoras postulated that prior to the creation event, the universe was a vast, amorphous mixture of a large, possibly infinite, number of fundamental, imperishable physical substances—the ingredients of all matter found throughout the cosmos today. He reasoned that these fundamental ingredients of matter, while ubiquitous, were not uniformly distributed and some were more plentiful than others.

Then at some point in time, the nous caused the vast primeval mixture to begin to rotate around a small point within it, and as the whirling motion expanded outward into the vast amorphous expanse, the ingredients in the mixture were separated in accordance with their differing densities. The separated ingredients remixed with each other in different places and in different ratios, ultimately producing the diversity of physical things, both animate and inanimate, that we perceive throughout the cosmos today.

Terminology aside, Anaxagoras' description of the origin of the universe has striking parallels with today's *big bang theory*, which postulates that, in the beginning, a pulse of energy appeared in what was then a void. The energy pulse began to manifest itself as radiation and expanded in all directions, eventually cooling enough for the basic building blocks of all matter, atoms, to form. Over time, as the universe continued to expand and cool, the atoms began to combine into molecules of matter, which eventually coalesced into galaxies, stars, and planets, creating the universe as we know it.

Anaxagoras did not propose a specific name for his fundamental entities and was somewhat vague as to their specific nature, but his contemporaries, Leucippus of Miletus and Democritus of Abdera, introduced the term *atomos* for the smallest possible unit of all matter. *Atomos* means indivisible in Greek and is the origin the scientific term *atom*. The concept of the atom being the smallest, indivisible, and imperishable particle of matter was a scientific axiom until the end of the nineteenth century when it was discovered that the atom was itself divisible and had even smaller component parts.

Anaxagoras proposed that an object's different features—colour, shape, size, and so on—could be explained by a unique set of characteristics that was a function of the dominant substance that constituted the object. Atomic theory today states that matter is composed of molecules, in turn composed of different combinations of atoms, and it is the molecules that impart the different physical attributes to objects.

Anaxagoras also postulated that the number of the fundamental substances was constant and unchanging in nature so that when objects such as living organisms decompose, the fundamental substances do not disappear but rather separate and recombine into other forms of matter. Today, this concept is embodied in the scientific law of conservation of mass.

According to Anaxagoras, the nous was responsible for the creation and functioning of the universe, and that the nous existed in a realm that was external to, and separate from, the fundamental substances that comprised the universe. Anaxagoras also postulated that it is the presence of the nous within living organisms that gives them consciousness and vitality.

To Anaxagoras, the nous was an impersonal, amoral, rational force that organized and maintained the ordered functioning of the cosmos. While some theologians equate Anaxagoras' nous with the Judaic God or to some Greek gods and goddesses, Anaxagoras himself never linked the nous with any supernatural deity, and he was considered by his contemporaries to be an atheist.

Centuries after Anaxagoras' death, several philosophers of classical Greece continued to advance his theories of matter and the concept of the nous as the creator of the world. Socrates, Plato, and Aristotle all expanded on his work and introduced the notion that because the wisdom of the nous was manifest in the natural world, it could be accessed by humankind through observation and logical reasoning. They believed that such reasoning would not only provide humankind with knowledge and understanding of the natural world, but would also inevitably lead to the realization that living a virtuous life, guided by moral and ethical principles, was the only logical choice. They argued that good and virtuous behaviour led to a life of harmony with one's fellow human beings and with the cosmos as a whole. And with such harmony came true happiness and inner peace.

Around 300 BCE, the Greek philosopher Zeno of Citium founded a school of philosophy known as *Stoicism*. Stoicism drew upon the ideas of previous Greek scholars to form a spiritual philosophy that formalized the principles of virtue ethics and presented an integrated theory of the nature of the cosmos and the human soul.

According to the Stoics, the cosmos in its entirety is a material, reasoning substance or pantheistic entity known as *God* or *Nature*. The Stoic god is neither transcendent, like Anaxagoras' nous, nor providential, like the Judeo-Christian god, but rather, the sole, immanent divine essence that constitutes the cosmos itself. In essence, the Stoics believed that god *is* the cosmos, and the cosmos is god, the only realm of existence. Hence, they did not believe in an afterlife as a reward for living a good and virtuous life, but rather, that life itself is its own reward.

The Stoic god consisted of two integrated substances: *matter*, the physical material of which the universe is composed, and the *Logos*, the universal consciousness or mind of god, the supreme reasoning power and the source of the laws and principles that underlie the evolution of the cosmos.

In the Stoic view of creation, god released a creative, animating substance, known as the *Pneuma*, upon the disorganized, inanimate world. The term *pneuma* was ancient Greek for the *breath of life* and was later translated into Latin as *spiritus*, the root of the English term *spirit*. The pneuma gave matter form and motion and organized the cosmos into four types of being—minerals or inanimate material, plants, non-rational animals, and humans—organized according to the *Scala Naturae* or *chain of being*. The Scala Naturae, shown in Table Four, is a hierarchical ordering of the four types of being according to the level of consciousness and mental capacity that was imbued in each by the pneuma.

TYPE OF BEING	FORM OF PNEUMA	CAPACITIES
God	Perfect Logos	Universal Consciousness, Vitality, Supreme Reason
Humans	Logos	Material Form, Consciousness, Vitality, a Soul and Reason
Non-rational animals	Psyche	Material Form, Consciousness, Vitality, and a Soul
Plants	Phusis	Material Form, Vitality
Minerals	Hexis	Material form

Table Four - The Scala Naturae of Stoicism

At the bottom of the chain are the minerals, or physical objects and materials that constitute the inanimate entities on earth. Next up the chain are the plants, which are imbued with the vitality of life but have no consciousness. Next are the animals, which have vitality, consciousness, and a *psyche*, or *soul*. Humans are positioned above all other animals in the chain of being in that they possess not only vitality, consciousness, and a soul, but also the rationality or reasoning powers of the logos. At the top of the chain is the creator, the universal consciousness, the source of vitality, and the perfect logos.

The Stoics believed that of all living organisms, only humans possess the logos. The unique human capacity for acquiring knowledge through logical thought, intuition, and imagination stems from the reasoning powers of the logos as manifest in the human mind. These reasoning powers are what enable humankind to acquire an understanding of the natural world and discover the divine wisdom and grandeur inherent in the workings of the cosmos. Drawing on the ideas put forth by Socrates, Plato, and

Aristotle, the Stoics believed that wisdom and virtue are the highest good and stem from having knowledge of the natural world acquired through reason. Such knowledge, they maintained, is what inspires humankind to live a good and virtuous life. The wise and virtuous live in harmony with the divine wisdom found in nature, which enables them to find happiness, purpose, and meaning in life.

To the Stoics, there was no providential god that would intervene in an individual's life in times of need, but rather, it was up to the individual to overcome hardship through reason, perseverance, and acceptance. It was the endurance of hardship and overcoming of adversity that gave humankind wisdom and character, strengthening their resolve to live a good and virtuous life. The term stoic is often used even today to describe a person who can endure pain or hardship without complaint.

Stoicism flourished throughout the Greek and Roman Empires until around the end the third century CE when Christianity began to gain favour as a religion in Rome. In 313 CE, the Roman Emperor Constantine legalized Christian worship with the Edict of Milan, and in 380 CE, under the emperor Theodosius I, Christianity became the official religion of the Roman Empire.

Stoicism as a spiritual philosophy is still held by many people today. Not surprisingly, many eminent scientists have been Stoics or held beliefs in closely related spiritual philosophies such as Pantheism. Carl Sagan, Albert Einstein, Nicolas Tesla, and Stephen Hawking were all professed pantheists.

Stoicism and the Judeo-Christian Religious Traditions

> *"In the beginning was the Word, and the Word was with God, and the Word was God. The same was in the beginning with God. All things were made through him; and without him was not anything made that hath been made. In him was life; and the life was the light of men."*
>
> *...The Christian Bible, King James Version, John 1:1 to 1:4*

The Stoic philosophy and the Judeo-Christian religious traditions all teach that a divine being or God created the cosmos and that the soul is a manifestation of that God within each of us. Each argues for a similar code of ethics and each teaches that a good and virtuous life should be the goal of all humankind.

The difference between the Stoic philosophical view and the Judeo-Christian religious traditions is fundamental, however. Whereas Stoics believe that God is immanent in the cosmos but does not intervene in its ongoing operation, Christians and Judaists believe that God is transcendent and providential, proactively guiding the ebb and flow of the cosmos and the course of individual human lives.

Stoics believe that access to the logos, or God's wisdom, is through reason, whereby we gain an understanding of the universe and learn to accept and deal with adversity on our own. Judaists and Christians, however, believe that God is accessible via prayer, meditation, and ritual, through which one can learn of God's will and appeal to God to intervene on one's behalf.

Whereas Stoics develop ethical principles through reasoning and scientific knowledge, Judaists obtain their moral guidance directly from God as laid down in the Hebrew Bible, and Christians gain similar guidance through the Old Testament and the teachings of Jesus Christ as written in the gospels of the New Testament.

Both Christians and Judaists believe that those of us who live a good and moral life are rewarded in the afterlife with some form of existence whereby our soul resides in *God's Kingdom of Heaven* for all eternity. Stoics do not believe the individual soul survives death, but rather that living a virtuous life is its own reward. They would argue that the concepts of heaven and hell relate to the choices humankind makes in the course of daily life and that through those choices, heaven or hell are created right here on Planet Earth.

While both Judaism and Christianity share the same ancient roots that predate Stoicism, Christianity appears to have been strongly influenced by Stoic philosophy. The four gospels of the New Testament—Matthew, Mark, Luke, and John—were originally written in Greek at the time that Stoicism was the predominant philosophy throughout the Roman Empire, which, by then, also included Greece. Hence, even though they were Jewish, Christ's disciples were most likely familiar with the concepts of Stoicism and, given that they appear to have been pre-disposed to consider alternatives to Judaism, may even have been students of the Stoic philosophy prior to meeting Jesus of Nazareth.

The King James version of the Christian Bible, cited above, was translated by English theologians in 1611, and the translation was based on various previous translations from the original Greek (in the case of the New Testament) and Hebrew (in the case of the old Testament). Nevertheless, the translation of John 1.1 to 1.4 strongly reflects the ancient Stoic philosophy. In the King James translation, the full meaning of John's text hinges on the underlying meaning of the word, *Word*, which is used three times in the first sentence. In the original Greek, the word *Logos* was used in each of these instances, and the Christian scholars that undertook the King James translation understood that John had used the word *Logos* to refer to Jesus Christ, the Son of God incarnate, sent to earth by God for the redemption of humankind. Hence, when John referred to Christ as the *Logos*, he was saying that Christ was the incarnation of the supreme or perfect Logos of God, which places Christ on a parallel with God in the Stoic Scala Naturae. As the son of God, Christ is given an elevated position relative to the rest of humankind, who are incarnations of a somewhat less than perfect form of the Logos. Such a notion reinforces the likelihood that John was strongly influenced by Stoic philosophy as the introductory passages of his gospel reflect essentially the Stoic view of god, creation, vitality (life), and the rational soul, expressed as the "light of men."

CHAPTER FOURTEEN
The Science of the Soul

"Science is not only compatible with spirituality; it is a profound source of spirituality."

...Carl Sagan

SCIENCE IS FUNDAMENTALLY the study of the interaction of energy and matter. Matter is found in a myriad of forms that derive from the particular molecules comprising them. Energy is characterized by the way it interacts with matter, causing material entities to move or be transformed from one form, size, or shape to another. The various types of matter–energy interaction include mechanical, electromagnetic, kinetic, potential, gravitational, thermal, chemical, and nuclear.

While much is known about the physical characteristics of matter and the fundamental particles that comprise it, nothing is known about the physical nature of energy itself. The classic theories of physics, such as Newton's laws of motion and universal gravitation, Einstein's theory of relativity, or Maxwell's theory of electromagnetism, each describe the way energy causes matter to behave, but there are no theories that describe the fundamental nature of energy in its own right.

Physicists define energy as the ability to do work, but defining something as an ability doesn't tell us anything about the nature of the entity itself. Energy is like consciousness and the vitality of life in that we experience it every day, but we can't describe it in terms of inherent physical attributes. We talk about and use energy in virtually all aspects of our daily routine, from describing how energetic we feel to the energy in gasoline that fuels our automobiles. We use energy to cook our food, heat our homes, and power our machines. We have a vast knowledge of what forms energy takes, what energy can do and how to use it to our benefit, but the presence of energy itself can be experienced only through its interaction with matter—that is, the myriad of roles energy plays in the physical world around us and within the bodies of living organisms. It seems ironic that energy, the very entity that lies at the heart of virtually all scientific theory, has no physical manifestation in itself, and nothing is known about its essential nature.

In 1905, Albert Einstein published his *theory of relativity*, which established the *principle of mass-energy equivalence*, stating that anything having mass also has an equivalent amount of energy associated with it. This principle is embodied in Einstein's famous equation:

$E = mc^2$

where **E** is energy in joules, **m** is mass in kilograms, and **c^2** is the velocity of light squared.

Einstein's equation provides a mathematical model defining how energy and mass are transformed from one to the other and how energy and mass are different manifestations of the same entity. The term **c^2** is an extremely large number, equal to 9×10^{16} (nine followed by 16 zeros), and the equation shows that an enormous amount of energy can be released with the conversion of a small amount of matter, a fact that was exploited in the development of both nuclear reactors and the nuclear bomb.

Even though they were published over one hundred years ago, Einstein's theory of relativity and the associated equations still provide the conceptual framework for studying some of the most awesome matter–energy transformations that are ongoing throughout the cosmos, in particular, the phenomena of *black holes* and *white holes*.

A black hole is defined as, "a region of space having a gravitational field so intense that no matter or radiation can escape it." Einstein predicted the existence of black holes in 1916, but it wasn't until 1971 that a black hole was first actually observed. Up until that discovery, few scientists believed Einstein's prediction, dismissing it as a physically baseless artifact of the mathematics of relativity. But since their discovery, black holes have become one of the most intensely studied phenomena in the field of cosmology, and much has been learned about their structure and behaviour.

It is now known that black holes typically form when a star collapses upon itself at the end of its life cycle. The matter from the dying star is crushed under gravitational forces, forming a *gravitational singularity*, an extremely dense cluster of mass with, according to Einstein's equation, an unfathomably large amount of associated energy. After a black hole has formed, it continues to grow by absorbing matter from its surroundings, including other stars and their planetary systems, even absorbing or merging with other black holes. Some black holes are incomprehensibly large, having estimated masses equivalent to millions of solar masses (one *solar mass* is equal to the mass of our sun, estimated at 1.989×10^{30} kilograms). As more matter is attracted and crushed into the central point of mass, the gravitational attraction of the black hole becomes enormous. So powerful is the gravitational "pull" of these mysterious cosmic entities that even light itself cannot escape them, hence their dark appearance and the term *black hole*.

The mathematical equations that predicted the existence of black holes also predict the existence of the opposite phenomenon, *white holes*. To date, no white holes have been observed, and their existence remains a mathematical prediction. Nevertheless, given the history of black hole prediction and discovery, some cosmologists believe white holes do exist and have developed theories concerning their physical characteristics and relationship with black holes.

One theory of white holes speculates that at some point in the life cycle of a black hole, a *big bang* explosion occurs at the gravitational singularity, causing the black hole to reverse itself and begin to spew forth radiation, transforming it into a white hole. In contrast to a black hole, a white hole exerts no gravitational pull, implying that the mass that was present in the gravitational singularity of the original black hole, has been converted into pure energy in the white hole. Electromagnetic radiation, including light, is subsequently expelled from the white hole, giving it a bright appearance, hence the term white hole. It is also speculated that as the radiation spews outward from a white hole, it expands and cools, allowing particles of matter to form, eventually coalescing into new galaxies of stars and planetary systems. Some cosmologists speculate that the initial explosive event, or big bang, that initiated the formation of the universe, may have been the result of a massive black hole reversing into an equally massive white hole.

The currently accepted big bang model of the origin of the universe postulates that, about 14 billion years ago, a pulse of energy appeared somewhere in the void that was to become space. The energy pulse itself was of essentially infinite intensity but was dimensionless as conditions were too hot for any matter to exist. Indeed, the concept of location and physical size had no basis because, without matter, the universe itself did not exist.

Other than the tenuous postulation that the creation event occurred within a white hole, science has no convincing theories of how, why, or from where the initial energy pulse originated. But in terms of understanding what happened following the big bang, remarkable advancements have been made.

Immediately following the appearance of the energy pulse, it began to manifest itself as radiation, which expanded outward in all directions, causing it to cool enough for matter to form. The matter that first formed existed for the tiniest fractions of a second and had almost no mass, behaving more like tiny packets of energy than matter as we know it. But over time, as the radiation further expanded and cooled, larger and more stable particles began to appear. Quarks and gluons formed, which then came together to form protons and neutrons, which combined with electrons to form the first atoms. As the atoms of all elements of the periodic table eventually formed, they began to come together as molecules of matter, and as trillions of molecules swarmed throughout the new universe, they coalesced under gravitational forces into ever-larger aggregates of matter, eventually forming the familiar galaxies of stars and planets observed today throughout the cosmos.

Black holes have been observed within most galaxies, slowly absorbing the stars and planetary systems in their vicinity. Such discoveries have prompted some cosmologists to theorize that entire galaxies may be gradually destroyed by black holes and then reformed when the black holes transform into white holes. And given that there are billions, if not trillions of galaxies throughout the universe, many may be in the process of death or rebirth at any point in time. The theory is pure conjecture at this point in time, and there are many competing theories being proposed to explain the phenomena of black and white holes, but the implications of most of these theories are similarly profound. If galaxies are continuously being destroyed and reformed, then the universe as we know it may not be the result of a single big bang event, but rather big bang events may be a frequent and ongoing occurrence. If so, then the universe is being continuously rejuvenated, and will likely continue to be so, for all eternity. The universe may have neither a beginning nor an end.

As we examine the universe today, we find that over 96 percent of what we see is empty space, filled with energy and matter in a variety of forms, some of which are still not entirely understood. The visible matter in the universe—the galaxies, stars, and planets—occupy less than 4 percent of the volume of the universe, and the remaining ninety-six percent is filled with energy, that same energy that appeared at the time of the big bang.

But what of matter itself?

Until the turn of the nineteenth century, it was the scientific view that all matter in the universe was composed of fundamental, indivisible, solid particles called atoms, a notion first introduced in ancient Greece by the natural philosophers, Leucippus and Democritus. But around the turn of the nineteenth century, physicists like Ernest Rutherford, Neils Bohr, and several others, revealed to the world that atoms were not the smallest particles of matter but rather they themselves were composed of even smaller particles. And over the course of the last one hundred years, much has been discovered about the internal structure of atoms and those components, or *sub-atomic particles*, that comprise them.

Today, the atom is modelled as a central *nucleus*, surrounded by an *energetic cloud* of one or more electrons orbiting about it at very high speed. The nucleus is composed of protons and neutrons and varies in size depending on the number of protons and neutrons it contains. But the average nucleus is about 10^{-9} microns, or 0.000000000000001 metres in diameter.

Electrons, on the other hand, are rather enigmatic entities, sometimes behaving like tiny particles of matter, but at other times, more like waves of energy. Because of this ambiguous behaviour, physicists use both an energetic wave model and a particle model interchangeably to explain the behaviour of electrons under different circumstances. Hence, the true nature of the electron is unknown, and its behaviour is yet to be completely explained in terms of a single physical model. In most current models of the atom, however, the electrons are viewed as an organized energy field surrounding a solid nucleus.

The diameter of an atom is defined as the diameter of the electron energy field, which varies with the number of electrons it contains, ranging between 1×10^{-4} and 5×10^{-4} microns. The nucleus occupies only 1.0×10^{-9} percent of a typical atom's volume, and the rest of the atom is filled with the electron energy field.

Hence, even the atom, the basic building block of all matter, is almost entirely energy and, this, combined with the scarcity of matter in the universe, implies that the universe itself is almost entirely energy. The entire cosmos, including both animate and inanimate entities, is imbued with energy, the same energy that mysteriously appeared in the void some fourteen billion years ago and transformed itself into the universe we see today. As Einstein and others have declared, "The cosmos is energy and energy is the cosmos!"

So minute is the actual amount of solid matter in the universe, comprised only of the nuclei of its constituent atoms, it would seem that matter serves only to provide the sheerest veils of material frameworks within which energy is organized into what we know as separate physical entities. Yet these diaphanous veils of matter are the means by which energy assumes different and separate physical forms, which in turn, exhibit different physical characteristics.

In spite of the extreme scarcity of actual matter, living organisms, including we humans, perceive only the material aspect of reality, and these perceptions provide the basis for our life experience. One might wonder how matter, and the myriad of physical entities it comprises, is what we perceive as representing reality, when the amount of it in material objects is so small and appears to exist only to organize energy into separate physical entities.

Perhaps reality isn't quite as it is perceived to be.

When we strike an iron nail with a hammer, perhaps to drive it into a piece of wood, we hear a resounding clang, and the nail is driven deep into the wood (or elsewhere!) Such an experience is about as undeniably physical or *real* as anything we do. In our mind's eye, we imagine that the atoms of the hammer and of the nail are tightly packed, so they each form an extremely hard surface. We perceive that it is the impact of these two solid surfaces striking each other that causes the noise and drives the nail.

This is not the case, however. In fact, it is the energy in the force field created by the electrons swirling about the iron molecules in both the hammer and the nail that repels the two, keeping them apart. When the electron energy field of the hammer and nail get close enough, the repulsive force is enormous, and it creates the sound and movement of the nail. In fact, the hammer never reaches the nail in a material sense, as the repulsive forces of the energy field surrounding the molecules in the two objects are so strong that the surfaces of the hammer and the nail never make contact. If they did, the nail wouldn't go anywhere as it would be so strongly fused with the hammer that they would both end up being parts of the same iron object.

From a scientific perspective, the cosmos is essentially organized energy. And it is that organized energy within living organisms that imparts the vitality of life and consciousness, and in the case of humans, a reasoning mind and an emotional soul.

But how is this scientific theory any different than the philosophical or religious perspectives that were conceived over two thousand years ago? Terminology aside, the scientific theory of creation and ancient philosophical and religious thought are remarkably similar. Certainly, scientific theory reflects a much more detailed understanding of creation, but at the conceptual level, there is little difference between the *Nous* of Anaxagoras, the *Logos* of Stoicism, the Judeo-Christian *God*, and the *Energy* of science. Each is deemed by its proponents to have created the cosmos through the organization of matter in a very similar fashion. And just as science has theorized that the same mass–energy transformation principles that governed the formation of the universe are what account for the life-sustaining vitality, consciousness and the soul within living things, Anaxagoras' *Nous*, the Stoic Logos, and the Judeo-Christian God are deemed to be the source of these same entities. It is also the Stoic view that God is immanent throughout the cosmos, that God and the universe are one and the same, a notion strongly supported by the modern scientific theory that the cosmos is essentially organized energy.

So what exactly *is* the Creator if it is not the energy that was concentrated in the initial impulse from which the universe sprung? And what *is* the act of creation if it is not that same energy organizing itself within veils of matter to form the cosmos we know today? What is the soul if it is not the energy that constitutes the vitality of all living things?

It seems compelling to accept the notion that what science calls *Energy*, Animists call the *Great Spirit*, Stoics call the *Logos*, Christians and Jews call *Jehovah* or *Yahweh*, and the followers of Islam call *Allah*. Science, and each of these faiths, may assign different names and attributes to their god, revere and celebrate it in different ways, but they are one and the same. They are the creator of the universe and the consciousness, vitality, and soul that reside in living things.

The Spiritual Essence of Science

"Quantum physics thus reveals a basic oneness of the universe."

...*Erwin Schrodinger*

The ancient Greeks developed a spiritual philosophy that taught that the reasoning powers of the logos were manifest in the human mind. These reasoning powers—conceptualization, logical thinking, intuition, and imagination—are what gives the mind of *Homo sapiens sapiens* its remarkable capabilities, allowing humans to develop an understanding of the natural world, to discover the virtue of living a moral and ethical life and perhaps ultimately reveal the true nature of the creator.

The scientific knowledge we possess today is the result of the application of a rigorous methodology of observation, inductive reasoning, and independent verification, known as the scientific method, an objective process of discovery that was first introduced in ancient Greece over two thousand years ago. The ancient Greeks applied the scientific method to gain an understanding of how the natural world operated, believing that such pursuits would lead to an appreciation for the wisdom that underlies the functioning of the cosmos, and provide the rationale for living a moral and virtuous life. Such knowledge and beliefs formed the basis for the Stoic philosophy that has strongly influenced western philosophical, religious, and ethical thought to this day. And in many ways, modern scientific theory is only providing us with more detailed explanations of concepts and principles first developed by the ancient Greek natural philosophers.

Perhaps, as the ancient Greeks believed, the ultimate purpose of science is to not just discover the laws and principles that underly the ongoing evolution of the cosmos, but to reveal the wisdom inherent in those principles and to perhaps even reveal the source of that wisdom, providing insight into the enduring questions of antiquity: how and why was the universe created, what is the nature of its creator, and what is the source of the vitality of life?

Many prominent scientists appear to have reflected on this ultimate goal of science.

The electrical engineer and physicist, Nikola Tesla, one of the all-time greatest pioneers in applied electricity, is quoted as saying, "My brain is only a receiver. In the Universe there is a core from which we obtain knowledge, strength and inspiration."

Albert Einstein, creator of the theory of general relativity, wrote, "That deep emotional conviction of the presence of a superior reasoning power, which is revealed in the incomprehensible universe, forms my idea of God."

And Stephen Hawking, eminent physicist and author of several books on relativity and cosmology, once remarked, "You cannot understand the glories of the universe without believing there is some Supreme Power behind it."

In ancient Mesopotamia, Greece, Egypt, and middle-aged Europe, there was little distinction made between scientific and religious knowledge.

Rather, they were regarded as equally valid parts of a single body of knowledge, and to a large extent, the role of scientific study was to confirm the truth of what religious scholars had known intuitively all along or to reveal the methodology that the creator had used in the formation and evolution of the universe. This combined knowledge discipline was referred to as natural philosophy, and throughout most of recorded history, many of the religious leaders in our societies were also the leading natural philosophers of their day.

However, as the scientific revolution began to gain momentum, and great scientists like Copernicus, Galileo, and Darwin made their discoveries about the functioning of the universe and the nature of human origins, they revealed truths that deeply troubled religious leaders. These and other scientists proved that Planet Earth was not the centre of the universe and that humankind did not necessarily enjoy a privileged position in the eyes of the creator. Such discoveries proved too much for most religious leaders to accept, and science and religion parted ways shortly after the scientific revolution began.

But today is an exciting time for science. New theories about energy and matter are emerging that are advancing our knowledge of these elusive entities. Quantum theory, for example, is profoundly changing the very way we view reality, affording physicists a theoretical framework that is bringing humankind very close to understanding the nature of the universe at the very instant of its formation. It is entirely possible that quantum theory may soon enable us to discover the true nature of energy itself, and in the process, reveal the nature of the creator and the source of the wisdom that underlies the awesome grandeur of the cosmos.

Sadly, however, at a time when it would appear that our scientists may be arriving at the very threshold of God's chambers, they are alone, long since abandoned by theologians and religious leaders. But it is not the goal of science to deny or refute our spirituality, but rather to strengthen it through knowledge and understanding. It is quite possible that our scientific knowledge may one day include a unified theory or belief system that not only reveals the true relationship between energy and matter, but also reveals the true nature of the Creator, of consciousness, the vitality of life and the soul.

CHAPTER FIFTEEN
The Lingering Question

"Treat the earth well: it was not given to you by your parents, it is loaned to you by your children. We do not inherit the earth from our ancestors, we borrow it from future generations."

...*Crazy Horse*

FOR MANY OF us, our intuition tells us that the creation event was the deliberate act of a supreme consciousness, with the intent to create a physical universe wherein all living organisms can enjoy a life experience.

But could it be that the universe was created for an even more profound purpose?

Perhaps the cosmos exists so that the Creator itself can experience life in a material form. At any given moment in time, there is an unfathomably large number of life forms on the earth, living out their lives in accordance with the capacities with which evolution has endowed them. Within each life form, the life experience varies in accordance with its level of consciousness and its physical and mental capabilities. If the Creator does indeed reside within all living things as many believe, then the Creator itself may be consciously sharing the life experiences of every living entity, ranging from the dim awareness of a single-celled organism to the incredible richness of human consciousness.

Since the first appearance of life on earth, some four billion years ago, and its subsequent evolution through countless numbers of different life forms, the Creator has experienced an incomprehensibly large number of diverse life experiences. And given that there is very likely life forms evolving on millions, if not billions, of planets throughout the universe, and if indeed the cosmos is continually being regenerated as some cosmologists now speculate, then perhaps the Creator has been experiencing life, and will continue to experience life, for all eternity.

Still, it must be acknowledged that some of us see no reason to associate deliberate intent with creation. To such individuals, the universe is not the handiwork of a conscious mind but rather simply the result of a random, fortuitous event. With such reasoning, there is no need to associate a meaning with life, as the big bang event that initiated formation of the universe occurred without conscious intent, and the laws and principles governing the universe are just the way energy and matter happen to behave.

So there remains the central, enduring question of antiquity, one that has divided humankind since some ancient member of our species first looked up at the night sky and wondered at the grandeur of it all. Why was the universe created? Is it the result of a chance event that just happened to occur without intent and for no purpose? Or was it an intentional act by a supreme consciousness for the purpose of creating a physical world where all life forms can have the experience of living in a material realm?

This lingering question is one that each of us must answer for ourselves, because the answer lies within. It lies in our soul and in the emotions we experience in the course of our daily lives. Every thoughtful person has at some point asked themselves this question and found the answer in their own way.

Many eminent scientists have found their answer in the beauty and elegance of the laws and principles that govern the operation of the cosmos. Those of faith have found their answer through prayer, meditation, and religious ritual.

Those of us who are fortunate enough to have children may have found our answer in the feeling of love that is evoked at the moment we first we look into the eyes of our newborn child. Others find their answer while listening to a favourite piece of music as it resonates with their soul and fills their heart with intense feelings of joy, sadness, or nostalgia.

Many of us find our answer in the natural world, in the beauty and fragrance of a rose as it opens to greet the rising sun on a summer morning. Or in the sound of a songbird, sitting in the rushes beside the stillness of an evening pond, singing, it seems, to celebrate the sheer joy of being alive.

There are countless sights and sounds that can stir our emotions and awaken an awareness of the Creator within us, reminding us of the exquisite beauty and awesome grandeur of the world in which we live. If you experience even one of life's countless such moments, you will have your answer and will have no doubt about the meaning of life and the reason our universe was created.

Perhaps, then, this is life's purpose. All living creatures are here simply to embrace and enjoy the experience of living, each in their own way, with the capacities endowed them by the Creator. And since the same divine essence of the Creator resides in all living things, it behooves every one of us—in fact, it is humankind's moral and ethical responsibility—to hold all life as sacred and treat all living creatures with the same love and respect with which we treat those persons nearest us.

And this same moral and ethical behaviour, this same love and respect, must also govern the way we treat our home, Planet Earth. We must remind ourselves how fortunate each of us is to be gifted a time on this beautiful planet, with the capacity to embrace and enjoy a life experience with such profound cognitive and emotional capacities.

But we must also accept the fact that with this gift of life comes a daunting, even terrifying, responsibility for humankind.

We humans are wreaking havoc on our home, Planet Earth. So successful have we been as a species that our numbers and demands are now too great for the earth to sustain, and we are exhausting our planet's ability to provide us with the resources we need to continue to behave as we have in the past. We are polluting the earth with toxins and waste, contaminating the very ecosystems that sustain us and irredeemably destroying the biomes that maintain the habitability of the planet.

So rapidly are we altering the earth that the natural restorative and evolutionary processes simply cannot operate within the accelerated time frames of the environmental changes we are causing. Only we humans, with our intellectual capacity, scientific knowledge, and technology, are in a position to arrest the catastrophic changes that now threaten our survival and the survival of much of the world's flora and fauna.

Humankind is now the custodian of Planet Earth, and responsibility for the well-being of our earth, and the future evolutionary direction of all the amazing life forms it supports, now rests solely with us. Surely we can find the will to put our religious and ideological differences aside and come together in a global, united effort to restore the earth to the magnificent and bountiful home it once was, so that our children, our grandchildren, and the many generations yet to come, will have the opportunity to enjoy our planet's amazing wonders the same way we have.

This is our destiny. It is the meaning of life.

- The End -

IMAGE CREDITS AND LICENSING

1. **Pictorial rendition of the Earth in Mollweide projection, as it might have appeared near the end of the Mesozoic era.** © 2016 Colorado Plateau Geosystems Inc.
2. **A 47 million year old prosimian fossil, *Darwinius masillae*, from Messel pit in Germany.** Credit: Jens L. Franzen, Philip D. Gingerich, Jörg Habersetzer1, Jørn H. Hurum, Wighart von Koenigswald, B. Holly Smith. Licensed under the Creative Commons Attribution-Share Alike 2.5 Generic Agreement
3. **A representation of *Proconsul africanus* based on a partially reconstructed skeleton on display in the Natural History Museum, Paris, France.** Image © Nobu Tamura - licensed under the Creative Commons Attribution-Share Alike 3.0 Agreement
4. **An artist's rendering of *Dryopithecus*.** Credit: Maxwell Swellteen, licensed under the Creative Commons Attribution-Share Alike License 3.0 Unported Agreement
5. **An artist's rendering of *Ardipithecus ramidus*.** © Jay H. Matternes. Reprinted by permission
6. **An artist's rendering of *Australopithecus Afarensis*.** Reprinted with permission from Encyclopædia Britannica, © 2005 by Encyclopædia Britannica, Inc.
7. **An artist's rendering of *Homo habilis*.** Reprinted with permission from Encyclopædia Britannica, © 2005 by Encyclopædia Britannica, Inc.
8. **A Stone Chopper fashioned in the Olduvan Tradition.** Credit: José-Manuel Benito Álvarez, University of Salamanca, Spain - licensed under the Creative Commons Attribution-Share Alike 2.5 Generic Agreement.
9. **Estimates of the Earth's average annual temperature over the last 500 million years, relative to the 1960-1990 global mean annual value, and including estimates for 2050 and 2100.** Compiled and published by Glen Fergus - licensed

under the Creative Commons Attribution-Share Alike 3.0 Unported Agreement.

10. **An artist's rendering of *Homo erectus*.** Reprinted with permission from Encyclopædia Britannica, © 2005 by Encyclopædia Britannica, Inc.
11. **A Hand Axe of the Acheulean Tradition.** Image © José-Manuel Benito Álvarez - Licensed under the Creative Commons Attribution-Share Alike 2.5 Generic Agreement.
12. **The Venus of Berekhat Ram.** Credit: Bahn P., 1998: The Cambridge illustrated history of prehistoric art, Cambridge University Press (Under License#15168).
13. **An artist's Rendering of *Homo sapiens neanderthalensis*.** Reprinted with permission from Encyclopædia Britannica, © 2005 by Encyclopædia Britannica, Inc.
14. **A Levallois Flint Point from Syria.** Credit: Guérin Nicolas - Licensed under the Creative Commons Attribution ShareAlike 3.0 Generic Agreement.
15. **The Reconstructed Skull of Omo 1 in the Natural History Museum, London.** © Credit: Natural History Museum, London / Science Photo Library. Under license.
16. **A comparison of a typical Modern Human skull (left) with that of a Neanderthal (right).** Credit: hairymuseummatt (original photo), Dr. Mike Baxter (derivative work) - Licensed under the Creative Commons Attribution ShareAlike 2.0 Generic Agreement.
17. **Robert Plutchik's Wheel of Emotions.** Public Domain. https://upload.wikimedia.org/wikipedia/commons/thumb/c/ce/Plutchik-wheel.svg/757px-Plutchik-wheel.svg.png.
18. **A Map of East Africa showing the Great Rift Valley, the Afar Triangle and the Nile River System.** Credit: Redgeographics - Licensed under the GNU Free Documentation License, version 1.2.
19. **Nassarius Shell Beads from Blombos Cave.** Image courtesy of Professor Christopher Hensilwood. Licensed by "Pictures in our Past".

20. **Still Bay Bifacial Stone Points found at Blombos Cave.** Credit: Vincent Moore, INRAP - Licensed under the Creative Commons Attribution ShareAlike 3.0 Unported Agreement.
21. **Engraved Ochre from Blombos Cave.** Credit: Henshilwood, C. S., d'Errico, F., and Watts, I. (2009). "Engraved ochres from the Middle Stone Age levels at Blombos cave, South Africa." Journal of Human Evolution, 57, 27–47. Licensed under the Creative Commons Attribution ShareAlike 4.0 International Agreement.
22. **Ostrich eggshell beads from Border Cave in South Africa.** Courtesy of Lucinda Backwell (with permission). Proceedings of the National Academy of Sciences
23. **The Löwenmensch.** Credit: Thilo Parg - Licensed under the Creative Commons Attribution ShareAlike 3.0.
24. **An example of Cave art from the limestone caves of Sulawesi.** Photo: Maxime Aubert, Griffith University (With Permission).
25. **The Top Predator of the Pleistocene, Arctodus simus, depicted beside a modern six foot male Homo sapiens sapiens.** Credit: Dantheman9758 - Licensed under the Creative Commons Attribution ShareAlike 3.0 Unported Agreement.
26. **The Lion Head found at Volgelherd Cave in the Swabian Jura.** Credit: Rainer Halama -Licensed under the Creative Commons Attribution-Share Alike 4.0 International Agreement.
27. **The Woolly Mammoth figurine found at Volgelherd Cave in the Swabian Jura.** Credit: Thilo Parg - Licensed under the Creative Commons Attribution-Share Alike 3.0 Unported Agreement.
28. **The Venus of Hohle Fels.** Credit Ramesis - Licensed under the Creative Commons Attribution-Share Alike 3.0 Unported Agreement.
29. **A bone flute found at Geißenklösterle Cave in the Swabian Jura.** Credit José-Manuel Benito - Licensed under the Creative Commons Attribution-Share Alike 2.5 Generic Agreement.
30. **Replica of the "Fighting Rhino and Horses Panel" from Chauvet Cave.** Credit HTO - Public domain, reference "https://en.wikipedia.org/wiki/Chauvet_Cave".

31. **Replica of the "The Panel of the Lions" from Chauvet Cave.** Credit HTO - Public domain, reference "https://en.wikipedia.org/wiki/Chauvet_Cave".
32. **The Ceramic Black Venus of Dolni Vestonice.** Credit Petr Novák, Wikipedia - Licensed under the Creative Commons Attribution-Share Alike 2.5 Generic Agreement.
33. **Solutrean Laurel Leaf points and tools.** Credit World Imaging, Wikipedia - Licensed under the Creative Commons Attribution-Share Alike 3.0 Unported Agreement.
34. **The oldest definitive evidence of the use of the Bow and Arrow from the Cova dels Cavalls in Spain.** Credit Joan Banjo - Licensed under the Creative Commons Attribution-Share Alike 3.0 Unported Agreement.
35. **Magdalenian tools and harpoon points.** Credit World Imaging - Licensed under the Creative Commons Attribution-Share Alike 3.0 Unported Agreement.
36. **A Magdalenian Carving of a Bison apparently licking an insect bite.** Credit Jochen Jahnke - Licensed under the Creative Commons Attribution-Share Alike 3.0 Unported Agreement.
37. **The Hall of the Bulls at Lascaux.** Credit Prof Saxx- Licensed under the Creative Commons Attribution-Share Alike 3.0 Unported Agreement.
38. **A Western Stemmed Point.** Credit (with permission) Pete Bostrom.
39. **A Bolen Projectile Point from Florida.** Reprinted (with permission) from the Peach State Archaeological Society Web Site, https://www.peachstatearchaeologicalsociety.org/index.php/2-uncategorised/131-bolen-bevel.
40. **The Iconic Clovis Projectile Point.** Image published by the Virginia Department of Historic Resources, https://commons.wikimedia.org/wiki/File:Clovis_Point.jpg.
41. **The Venus de Milo.** Credit Livioandronico2013, Licensed under the Creative Commons Attribution-Share Alike 4.0 International Agreement.

42. **Leonardo de Vinci's Vitruvian man.** Public Domain: https://commons.wikimedia.org/wiki/File:Vitruvian.jpg

BIBLIOGRAPHY AND RELATED READING

Earth History, Climate and Evolutionary Theory

1. Redfern, M. (2003). *The Earth: A Very Short Introduction,* Oxford University Press.
2. Darwin, Charles R. (1859). *On the Origin of Species by Means of Natural Selection, or the Preservation of Favoured Races in the Struggle for Life.*
3. Darwin, Charles R. (1871). *The Descent of Man, and Selection in Relation to Sex*
4. Darwin, Charles R. (1845). *Voyage of the HMS Beagle.*
5. Ruddiman, W.F. (2007). *Earth's Climate: Past and Future,* W. H. Freeman and Company, Macmillan Publishers.
6. Molnar, P. (2015). *Plate Tectonics: A Very Short Introduction,* Oxford University Press.
7. Brannen, P. (2017). *The Ends of the World: Volcanic Apocalypses, Lethal Oceans, and Our Quest to Understand Earth's Past Mass Extinctions,* Amazon Books
8. Schulte, P. (2010). *The Chicxulub Asteroid Impact and Mass Extinction at the Cretaceous-Paleogene Boundary.* Science Magazine, **327** (5970): 1214–1218
9. Furley, P.A. (2016). *Savannas: A Very Short Introduction,* Oxford University Press.
10. Charlesworth, B., and Charlesworth, D. (2017). *Evolution: A Very Short Introduction,* Oxford University Press.
11. Maslin, M. (2014). *Climate Change: A Very Short Introduction,* Oxford University Press.
12. Ellis, E.C. (2018). *Anthropocene: A Very Short Introduction,* Oxford University Press.
13. Annenberg Learner. *The Habitable Planet: A Systems Approach to Environmental Science,* https://www.learner.org/series/

the-habitable-planet-a-systems-approach-to-environmental-science/

Paleontology, Archaeology and Anthropology

14. Dunbar, R. (2004). *The World from beginnings to 4000 BCE, The Human Story - A new history of mankind's evolution,* Faber and Faber Ltd.
15. Boyd, R., and Silk, J. (2009). *How Humans Evolved, 5th Edition,* Norton, New York.
16. Smithsonian National Museum of Natural History, Human Origins Program. (2016). *What does it mean to be human?, http:// humanorigins.si.edu/evidence/behavior/stone-tools,*
17. River, Charles. (2018). *Homo Erectus: The History of the Archaic Humans Who Left Africa and Formed the First Hunter-Gatherer Societies,* Amazon Books.
18. Stringer, C. (2013). *Lone Survivors: How we came to be the Only Humans on Earth,* Times Books, Henry Holt and Company.
19. Young, J. (2019). *Neanderthals,* Amazon Books
20. Harari, Y.N. (2014). *Sapiens: A Brief History of Humankind,* McClelland & Stewart.
21. Bahn, P.G. (2012). *Archaeology: A Very Short Introduction,* Oxford University Press.
22. Parker, S. (2015). *Evolution: The Whole Story.* Thames & Hudson
23. Seddon, C. (2015). *Humans: From the beginning.* Glanville Publications
24. Roberts, A. (2018). *The Incredible Human Journey.* Bloomsbury.
25. Roberts, A. *(2010). Evolution: The Human Story, Second Edition.* Penguin Random House.
26. Wood, B. (2019). *Human Evolution, Second Edition: A Very Short Introduction,* Oxford University Press.
27. Bradshaw, J.L. (1997). Human Evolution: A Neuropsychological Perspective, Psychology Press
28. Roberts, M., and Pitts, M. (2013). *A Fairweather Eden: Life in Britain Half a Million Years Ago as Revealed by Excavations at Boxgrove,* Cornerstone Digital.

29. New World Encyclopedia. (2013). *http://www.newworldencyclopedia.org/entry/Bushmen.*
30. Rybin, E. (2009). *Tolbaga: Upper Paleolithic settlement patterns in the Trans-Baikal Region,* Archaeology Ethnology and Anthropology of Eurasia.

Psychology, Philosophy and Prehistoric Art

31. Bahn, P.G. (1998). *Cambridge Illustrated History of Prehistoric Art,* Cambridge University Press.
32. Anaxagoras (Author), Anderson, T. (Editor), Burnet, J. Translator). *Fragments of Anaxagoras,* Amazon Books.
33. Curd, P. (2019). *Anaxagoras,* Stanford Encyclopedia of Philosophy, https://plato.stanford.edu/entries/anaxagoras/
34. Malpas, J. (2012). "Donald Davidson", The Stanford Encyclopedia of Philosophy (Winter 2012 Edition), Edward N. Zalta (ed.), URL = <https://plato.stanford.edu/archives/win2012/entries/davidson/>.
35. Cooper, J.M. (editor). (1997). *Plato: Complete Works,* Hackett Publishing Company..
36. Barnes, J. (2000). *Aristotle: A Very Short Introduction,* Oxford University Press.
37. Inwood, B. (2018). *Stocism: A Very Short Introduction,* Oxford University Press.
38. Goenaga, L.O. (2008). *Jewish Stoicism: Analyzing the Evidence of Stoic Influence in Judaic-Christian Thought.* https://leonardooh.wordpress.com/2008/10/01/'jewish-stoicism'-analyzing-the-evidence-of-stoic-influence-in-judaic-christian-thought/
39. Crisp, R. (2015). *Social Psychology: A Very Short Introduction,* Oxford University Press.
40. Brooks, D. (2011). *The Social Animal: The Hidden Sources of Love, Character, and Achievement,* Random House
41. Duggan, W. (2007). *Strategic Intuition: The Creative Spark in Human Achievement,* Columbia Business School Publishing, Columbia University Press.

42. Evans, Jonathan St B. T. (2017). *Thinking and Reasoning: A Very Short Introduction*, Oxford University Press.
43. Bayne, T. (2013). *Thought: A Very Short Introduction*, Oxford University Press.
44. Passingham, R. (2016). *Cognitive Neuroscience: A Very Short Introduction*, Oxford University Press.
45. Evans, D. (2019). Emotion*: A Very Short Introduction*, Oxford University Press.
46. Strohm, P. (2011). *Conscience: A Very Short Introduction*, Oxford University Press.
47. Blackmore, S. (2018). *Consciousness: A Very Short Introduction*, Oxford University Press.
48. Margulis, E.H. (2018). *The Psychology of Music: A Very Short Introduction*, Oxford University Press.
49. Sarton, G. (1993). *Ancient Science through the Golden Age of Greece*, Dover Publications, Inc.
50. Plutchik, R. (2002). *Emotions and Life: Perspectives from Psychology, Biology and Evolution*, Academic Press Inc.

Religion and Spirituality

51. Dixon, T. (2008). *Science and Religion: A Very Short Introduction*, Oxford University Press.
52. Sheldrake, P. (2012). *Spirituality: A Very Short Introduction*, Oxford University Press.
53. Harvey, G. (2017). *Animism*, C. Hurst& Co.
54. Edward, C. (2018). *Animism: The Seed of Religion*, Createspace, North Charleston, SC.
55. Picton, A. (2016). *Pantheism: A Natural History*, Amazon Books, David Lane, Editor.
56. Harner, M. (1990). *The Way of the Shaman*, HarperSanFrancisco, A Division of Harper Collins Publishers.
57. Keown, D. (2013). *Buddhism: A Very Short Introduction*, Oxford University Press.
58. Woodhead, L. (2014). *Christianity: A Very Short Introduction*, Oxford University Press.

59. Bowker, J. (2014). *God: A Very Short Introduction,* Oxford University Press.
60. Scruton, R. (2002). *Spinoza: A Very Short Introduction,* Oxford University Press
61. Friedmann, D., and Sheldon, D. (2019). *The Biblical Clock: The Untold Secrets Linking the Universe and Humanity with God's Plan,* Amazon Kindle edition

Genetics and Biology

62. Slack, J. (2014). *Genes: A Very Short Introduction,* Oxford University Press.
63. Neal, J.M. (2016). *How the Endocrine System Works,* Wiley Blackwell Publishing
64. Barrett, L.F. (2017). *How Emotions are Made,* Houghton Mifflin Harcourt.
65. O'Shea, M. (2006). *The Brain: A Very Short Introduction,* Oxford University Press.
66. Allen, T., and Cowling, G. (2011). *The Cell: A Very Short Introduction,* Oxford University Press.
67. Cann, R.L., Stoneking, M., and Wilson, A.C. (1987). *Mitochondrial DNA and human evolution. Nature* 325:31–36

Cosmology and Physics

68. Greene, B. (2004). *The Fabric of the Cosmos - Space, Time and the Texture of Reality,* Vintage Books, A Division of Random House, Inc.
69. Chopra, D., Penrose, R., Stapp, H.P., Hameroff, S., Kafatos, M., Tanzi, R.E., Kak, S., King, C., Christensen Jr, W.J., Clarke, C.J.S., Nelson, R., Liljenstrom, H., Smythies, J., Nani, A., Cavanna, A.E., Furey, J.T., Fortunato, V. J., Bodovitz, S., Jansen, F. K., Martin, F., Carminati, F., Carminati, G.G., Crater, H.W., McDaniel, S.V., Carter, B., Mensky, M.B., Globus, G., Dobyns, Y.H., Gao, S., Kuttner, F., Rosenblum, B., Nauenberg, M., Vannini, A., Di Corpo, U. (2015, 2016). *How Consciousness Became the Universe,* Cosmology Science Publishers, Cambridge.

70. Polkinghorne, J. (2002). *Quantum Theory: A Very Short Introduction,* Oxford University Press.
71. Stannard, R. (2002). *Relativity: A Very Short Introduction,* Oxford University Press.
72. Hawking, S., and Mlodinow, L. (2010). *The Grand Design,* Bantam Books,
73. Close, F. (2009). *Nothing: A Very Short Introduction,* Oxford University Press.
74. Coles, P. (2001). *Cosmology: A Very Short Introduction,* Oxford University Press.
75. Carroll, S. (2017). *The Big Picture: On the Origins of Life, Meaning and the Universe Itself,* Dutton/Penguin Random House.

Milton Keynes UK
Ingram Content Group UK Ltd.
UKHW021326061124
2642UKWH00035B/159